T0419744

ADVANCES IN MATHEMATICS RESEARCH

ADVANCES IN MATHEMATICS RESEARCH

VOLUME 23

ADVANCES IN MATHEMATICS RESEARCH

ADVANCES IN MATHEMATICS RESEARCH

ADVANCES IN MATHEMATICS RESEARCH

VOLUME 23

ALBERT R. BASWELL
EDITOR

NOTICE TO THE READER

Library of Congress Cataloging-in-Publication Data

ISBN: 978-1-53612-512-2
ISSN: 1546-2102

Published by Nova Science Publishers, Inc. † New York

CONTENTS

PREFACE

In the opening chapter by Victor Martinez-Luaces, two kinds of matrices related to chemical problems are examined and an outline of their main properties about their eigenvalues is exhibited in order to demonstrate that all the ODE solutions are either stable or asymptotically stable. In chapter two by Ivan Kyrchei, the Cramer rules for the weighted Moore-Penrose solutions of left and right systems of quaternion linear equations are obtained. Next, in chapter three, Tadeusz Antczak showcases numerous sets of saddle point criteria for a new class of nonconvex nonsmooth discrete minimax fractional programing problems. Marcia de F. B. Binelo, Airam T. Z. R. Sausen, Paulo S. Sausen, and Manuel O. Binelo provide a summary of electric mathematical models used for the prediction of batteries charge and discharge behavior in chapter four. In chapter five, general methodology for the precise modeling and performance assessment of launch vehicles dedicated to microsatellites is proposed by M. Pontani, M. Palloney, and P. Teofilattoz. In chapter six, Nodari Vakhania exemplifies ties and relationships among some optimization problems such as scheduling and transportation issues. In chapter seven, a geometry without using points in established by N. L. Bushwick, bringing the book to a close.

In Chapter 1, two kinds of matrices related to chemical problems are analyzed. Firstly, the focus will be put on first order chemical kinetics mechanisms (FOCKM), which are modeled through ODE linear systems, where their associated matrices (FOCKM-matrices) have a particular structure. A summary of the main properties of their eigenvalues will be discussed in this Chapter. Taking into account these results it is possible to prove that all the ODE solutions are stable or asymptotically stable.

The second class of problems to consider are mixing problems (MP), also analyzed in previous works. These problems led to linear ODE systems, for which the associated matrices (MP-matrices) have different structures depending on whether or not there is recirculation of fluids. It can be observed that all the matrix eigenvalues have a non-positive real part and if the mixing problem involves three or less components, then all the eigenvalues have a negative real part and so, the corresponding ODE solutions are asymptotically stable.

Both types of matrices (FOCKM and MP matrices) have similarities and differences and the latter are important enough to obtain different qualitative behaviors of the ODE solutions as analyzed in Chapter 1.

The theory of noncommutative column-row determinants (previously introduced by the author) is extended to determinantal representations of the weighted Moore-Penrose inverse over the quaternion skew field in Chapter 2. To begin with, the authors introduce the weighted singular value decomposition (WSVD) of a quaternion matrix.

Similarly as the singular value decomposition can be used for expressing the Moore-Penrose inverse, Chapter 2 gives the representation of the weighted Moore-Penrose inverse by WSVD. Using this representation, limit and determinantal representations of the weighted Moore-Penrose inverse of a quaternion matrix are derived within the framework of the theory of column-row determinants. By using the obtained analogs of the adjoint matrix, the authors get the Cramer rules for the weighted Moore-Penrose solutions of left and right systems of quaternion linear equations, and for solutions of two-sided restricted quaternion matrix equation in all cases with respect to weighted matrices.

Numerical examples to illustrate the main results are given.

In Chapter 3, the authors present several sets of saddle point criteria for a new class of nonconvex nonsmooth discrete minimax fractional programming problems in which the involving functions are (Ψ, Φ, ρ)-univex and/or (Ψ, Φ, ρ)-pseudounivex. The results extend and generalize the corresponding results established earlier in the literature for such nonsmooth optimization problems.

Battery behavior modeling, under different use conditions, can be relatively complex due to the nonlinear nature of the charge and discharge processes. Understanding these dynamics by leveraging mathematical models, favors the development of more efficient batteries and also provides tools for software developers to better manage device resources. A review of electrical mathematical models used in the prediction of battery charge and discharge behavior is presented in Chapter 4. The class of electrical models has been used in various battery modeling applications, including mobile devices and electrical vehicles. The scientific investigation of such models is motivated by their capacity to provide important electrical information such as current, voltage, state of charge, and also some nonlinear aspects of the problem while keeping a relatively low complexity. Six subclasses of electrical models (Simple models, Thévenin-based models, Impedance-based models, Runtime-based models, Combined models and Generic models) along with a discussion of the main characteristics of each. This will demonstrate the evolution of electrical models through successive modification and combination, resulting in varying levels of accuracy and complexity.

Multistage launch vehicles of reduced size, such as "Super Strypi" or "Sword", are currently investigated for the purpose of providing launch opportunities for microsatellites. Chapter 5 proposes a general methodology for the accurate modeling and performance evaluation of launch vehicles dedicated to microsatellites. For illustrative purposes, the approach at hand is applied to the Scout rocket, a micro-launcher used in the past. Aerodynamics and propulsion are modeled with high fidelity through interpolation of available data. Unlike the original Scout, the terminal optimal ascent path is determined for the upper stage, using a firework algorithm in conjunction with the Euler-Lagrange equations and the Pontryagin minimum principle. Firework algorithms represent a recently-introduced heuristic technique, not requiring any starting guess and inspired by the firework explosions in the night sky. The numerically results prove that this methodology is easy-to-implement, robust, precise and computationally effective, although it uses an accurate aerodynamic and propulsive model.

Scheduling and transportation problems are important real-life problems having a wide range of applications in production process, computer systems and routing optimization when the goods are to be distributed to the customers using scarce available resources. In Chapter 6, the authors illustrate ties and relationships among some of these optimization problems. They consider scheduling problem with release and due dates, batch scheduling and vehicle routing problems. As the authors will show here, although these problems seem to have a little in common, a closer look at their parametric and structural properties can give us more insight into the “hidden” ties among these problems that may lead to efficient solution methods.

The construction presented in Chapter 7, like systems of Aristotelian continua presented elsewhere, is designed to establish a geometry without using points. However, it goes further in that the foundation consists of elements that are completely abstract, rather than line segments whose universe is a line. Furthermore, the result could be modified to represent elements and spaces of multiple dimensions.

In: Advances in Mathematics Research
Editor: Albert R. Baswell

ISBN: 978-1-53612-512-2

Chapter 1

MATRICES IN CHEMICAL PROBLEMS: CHARACTERIZATION, PROPERTIES AND CONSEQUENCES ABOUT THE STABILITY OF ODE SYSTEMS

Victor Martinez-Luaces
Electrochemistry Multidisciplinary Research Group,
Faculty of Engineering, UdelaR, Montevideo, Uruguay

Abstract

In this chapter, two kinds of matrices related to chemical problems are analyzed. Firstly, the focus will be put on first order chemical kinetics mechanisms (FOCKM), which are modeled through ODE linear systems, where their associated matrices (FOCKM-matrices) have a particular structure. A summary of the main properties of their eigenvalues will be discussed here. Taking into account these results it is possible to prove that all the ODE solutions are stable or asymptotically stable.

The second class of problems to consider are mixing problems (MP), also analyzed in previous works. These problems led to linear ODE systems, for which the associated matrices (MP-matrices) have different structures depending on whether or not there is recirculation of fluids. It can be observed that all the matrix eigenvalues have a non-positive real part and if the mixing problem involves three or less components, then all the eigenvalues have a negative real part and so, the corresponding ODE solutions are asymptotically stable.

Both types of matrices (FOCKM and MP matrices) have similarities and differences and the latter are important enough to obtain different qualitative behaviors of the ODE solutions as analyzed in the chapter.

1. Introduction

Several classical books show the strong relation between chemistry and mathematics, particularly in differential equations, Laplace transform, statistics and numerical methods [1, 2]. This relation was explored in previous papers and book chapters, including five books released by NOVA Publishers [3, 4, 5, 6, 7].

The mathematical tools used in these works included Laplace transform [6, 7, 8, 9, 10], Fourier transform [10, 11], ODE linear systems [5, 12, 13], parabolic PDE [6, 7, 10], numerical methods [14, 15, 16], non parametric statistics [17, 18] and experimental design [19, 20, 21], etc.

The connection of chemical problems with linear algebra was examined in previous works in two different directions. Firstly, chemical kinetics mechanisms were analyzed, with a special interest in first order reactions (called FOCKM-problems). Those FOCKM-problems and the corresponding FOCKM-matrices were the main subject in many papers and book chapters [3, 5, 13, 22, 23]. Also, another area explored in previous papers [12, 24] refers to mixing problems (MP), that can be modeled using linear ODE systems and for which the associated matrices (MP-matrices) have different structures, depending on whether or not there is recirculation of fluids.

Here, both kind of matrices – FOCKM and MP – are analyzed, including a characterization of each class (see section 2 and 4), followed by a study of some of their properties (see section 3 and 5).

In the last section their similarities and differences will be commented on, focusing on ODE solutions corresponding to both FOCKM and MP. In particular, the qualitative behavior of these solutions will be the main interest of the conclusion section.

2. Characterization of FOCKM-Matrices

A typical problem that involves first order ODE appears when a first order unimolecular chemical reaction is studied. This mathematical model is usually included in classical textbooks such as Courant [25], who adapted the original paper written by Wilhelmy in 1850 [26].

In a simplified version, a first order reaction can be easily schematized as follows: $A \xrightarrow{k} B$ where A represents reactant, k is the kinetic constant and B is the reaction product.

The corresponding ODE mathematical model for this chemical kinetics problem is:

$$\frac{d[A]}{dt} = -k[A], \tag{1}$$

where $[A]$ represents the concentration of substance, t is time and the negative sign indicates that the reactant is being transformed, so its concentration diminishes with time.

When several substances react and all these reactions are first order ones, we will call this system a FOCKM. The main purpose of this section is to obtain a general mathematical model (i.e., an ODE system) that describes accurately the FOCKM-problem. For this purpose, let us consider n chemical species E_1, E_2,..., E_n and suppose that all the possible first order chemical reactions $E_i \xrightarrow{k_{ij}} E_j$ take place. If any of these reactions does not occur, then, the corresponding kinetic constant will be considered null.

In this situation, direct reactions involving species E_1 are:

$$E_1 \xrightarrow{k_{12}} E_2,\ E_1 \xrightarrow{k_{13}} E_3, \ldots, E_1 \xrightarrow{k_{1n}} E_n \tag{2}$$

and the opposed reactions are:

$$E_2 \xrightarrow{k_{21}} E_1,\ E_3 \xrightarrow{k_{31}} E_1, \ldots, E_n \xrightarrow{k_{n1}} E_1. \tag{3}$$

Consequently, the corresponding ODE for the variation of E_1 concentration with time is:

$$\frac{d[E_1]}{dt} = -k_{12}[E_1] - k_{13}[E_1] - \ldots - k_{1n}[E_1] + k_{21}[E_2] + k_{31}[E_3] + \ldots + k_{n1}[E_n] \tag{4}$$

or in a condensed form:

$$\frac{d[E_1]}{dt} = -s_1[E_1] + k_{21}[E_2] + k_{31}[E_3] + \ldots + k_{n1}[E_n] \quad (5)$$

where

$$s_1 = k_{12} + k_{13} + \ldots + k_{1n} = \sum_{j \neq 1} k_{1j} \quad (6)$$

Following a similar reasoning for species E_i we have:

$$\frac{d[E_i]}{dt} = -s_i[E_i] + k_{1i}[E_1] + \ldots + k_{i-1,i}[E_{i-1}] + k_{i+1,i}[E_{i+1}] \ldots + k_{ni}[E_n] \quad (7)$$

where

$$s_i = k_{i1} + \ldots + k_{i,i-1} + k_{i,i+1} \ldots + k_{in} = \sum_{j \neq i} k_{ij} \quad (8)$$

Then, the ODE system that corresponds to a general FOCKM is:

$$\begin{pmatrix} [E_1] \\ [E_2] \\ \vdots \\ [E_n] \end{pmatrix}' = \begin{pmatrix} -s_1 & k_{21} \cdots & k_{n1} \\ k_{12} & -s_2 \cdots & k_{n2} \\ \vdots & \vdots & \vdots \\ k_{1n} & k_{2n} \cdots & -s_n \end{pmatrix} \begin{pmatrix} [E_1] \\ [E_2] \\ \vdots \\ [E_n] \end{pmatrix} \quad (9)$$

and its associated matrix is:

$$A = \begin{pmatrix} -s_1 & k_{21} \cdots & k_{n1} \\ k_{12} & -s_2 \cdots & k_{n2} \\ \vdots & \vdots & \vdots \\ k_{1n} & k_{2n} \cdots & -s_n \end{pmatrix} \quad (10)$$

where all the non-diagonal entries are non-negative and the diagonal elements are $-s_i = -\sum_{j \neq i} k_{ij}$, so all the matrix columns add up to zero.

To summarize, the characteristics of a FOCKM-matrix, we can give the following definition:

Definition 1. **A** is a FOCKM-matrix if the following conditions are satisfied:

A is a $n \times n$ matrix with real entries a_{ij}.

The non-diagonal entries are always non-negative numbers, i.e., $a_{ij} \geq 0 \quad \forall i \neq j$.

The diagonal elements are $-s_i$, where s_i is the sum of the non-diagonal entries in the i -th column.

To conclude, if all the reactions involved in the FOCKM are reversible, then:

$$E_i \xrightarrow{k_{ij}} E_j \text{ , with } k_{ij} \neq 0 \tag{11}$$

and

$$E_j \xrightarrow{k_{ji}} E_i \text{ , with } k_{ji} \neq 0 \tag{12}$$

So, in this particular case, all the non-diagonal entries are positive and all the diagonal ones are negative. These FOCKM-matrices corresponding to reversible reactions were studied in a previous paper [22].

3. Properties of FOCKM-Matrices

Taking into account the particular structure of these FOCKM-matrices, the following result can be stated:

Proposition 1. If A is a $n \times n$ FOCKM-matrix with $n \geq 2$ (i.e., two or more substances are involved in the mechanism), then $\det(A) = 0$.

Proof. If the three conditions of Definition 1 are considered, it is easy to observe that in a FOCKM-matrix $row_1 + row_2 + \ldots + row_n = \vec{0}$, and so the corresponding determinant must be zero.

Corollary 1. If A is a $n \times n$ FOCKM-matrix with $n \geq 2$ then $\lambda = 0$ is an eigenvalue of A.

This second result is an obvious consequence of having a null determinant.

In order to get relevant information about the non-null eigenvalues, let us consider $\vec{X} = (x_1, x_2, \ldots, x_n) \neq \vec{0}$ an eigenvector of A associated to the eigenvalue λ, then $(A - \lambda I)\vec{X} = \vec{0}$.

This equation can be written as:

$$\begin{pmatrix} -s_1 - \lambda & k_{21} \cdots & k_{n1} \\ k_{12} & -s_2 - \lambda \cdots & k_{n2} \\ \vdots & \vdots & \vdots \\ k_{1n} & k_{2n} \cdots & -s_n - \lambda \end{pmatrix} \begin{pmatrix} x_1 \\ x_2 \\ \vdots \\ x_n \end{pmatrix} = \begin{pmatrix} 0 \\ 0 \\ \vdots \\ 0 \end{pmatrix} \tag{13}$$

If we choose i such that x_i is the largest – in modulus – number in the vector $\vec{X} = (x_1, x_2, \ldots, x_n)$, the i-th product will be:

$$k_{i1}x_1 + k_{i2}x_2 + \ldots + (-s_i - \lambda)x_i + \ldots + k_{ni}x_n = 0 \tag{14}$$

Then

$$k_{i1}x_1 + k_{i2}x_2 + \ldots + k_{i-1}x_{i-1} + k_{i+1}x_{i+1} \ldots + k_{ni}x_n = (s_i + \lambda)x_i$$

and taking modulus, the following equation holds:

$$|s_i + \lambda||x_i| \leq k_{i1}|x_1| + k_{i2}|x_2| + \ldots + k_{i-1}|x_{i-1}| + k_{i+1}|x_{i+1}| \ldots + k_{ni}|x_n| \tag{15}$$

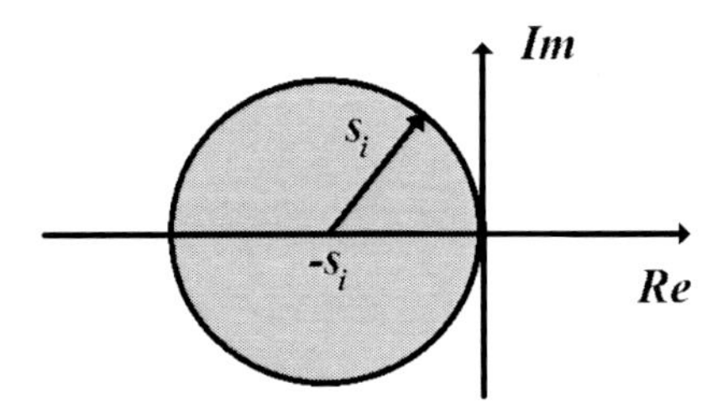

Figure 1. The circle within which the eigenvalue λ lies.

Taking into consideration that $|x_i| \neq 0$, since it is the largest number in the vector $\vec{X} = (x_1, x_2, \ldots, x_n)$, we have:

$$|s_i + \lambda| \leq k_{i1} \frac{|x_1|}{|x_i|} + k_{i2} \frac{|x_2|}{|x_i|} + \ldots + k_{i-1} \frac{|x_{i-1}|}{|x_i|} + k_{i+1} \frac{|x_{i+1}|}{|x_i|} \ldots + k_{ni} \frac{|x_n|}{|x_i|} \leq \tag{16}$$
$$\leq k_{i1} + k_{i2} + \ldots + k_{i-1} + k_{i+1} \ldots + k_{ni} = s_i$$

The final result is: $|s_i + \lambda| \leq s_i$ or: $|\lambda - (-s_i)| \leq s_i$ and this inequality means that λ lies in the circle centered at $-s_i$ with radius s_i , as shown in Figure 1.

This result was proved adapting the demonstration of the Gershgorin circle theorem, the first version of which was published by S. A. Gershgorin in 1931 [27]. This circle theorem may be used to find the bounds of the spectrum of a complex $n \times n$ matrix and its statement is the following:

Theorem (Gershgorin). If A is a $n \times n$ matrix, a_{ij} with $i, j \in \{1, \ldots, n\}$ are the matrix entries and $R_i = -\sum_{j \neq i} |a_{ij}|$ is the sum of the non-diagonal entries modules in the i -th row, then every eigenvalue of A lies within at least one of the closed discs $\overline{D}(a_{ii}, R_i)$.

Note 1. The closed disc centered at a_{ij} with radius R_i , i.e., $\overline{D}(a_{ii}, R_i)$, is called a Gershgorin disc.

Note 2. An obvious corollary of this result can be obtained by applying the Gershgorin circle theorem to A^T. The straightforward conclusion is that all the eigenvalues of A lie within the Gershgorin discs corresponding to the columns of A.

Application. As already observed, given a general FOCKM involving n species E_1, E_2 ... E_n the ODE system is like (9) and its associated matrix is given by (10). For instance, for the first column the Gershgorin disc is $\overline{D}(-s_1, R_1)$ where $R_1 = \sum_{j \neq 1} |k_{1j}| = \sum_{j \neq 1} k_{1j} = s_1$. Then, the Gershgorin disc corresponding to the first column is $\overline{D}(-s_1, s_1)$ and the same happens with all the other columns of A (10), so all the eigenvalues lie in $\bigcup_{i=1}^{n} \overline{D}(-s_i, s_i)$

It is important to note that every disc is centered on a non-positive number $-s_i \leq 0$ and the circle radius is the absolute value of this number (i.e., $|-s_i| = s_i$), so that all the Gershgorin discs are tangent to the imaginary axis. So, a given eigenvalue will lie in a circle like the one schematized in Figure 1.

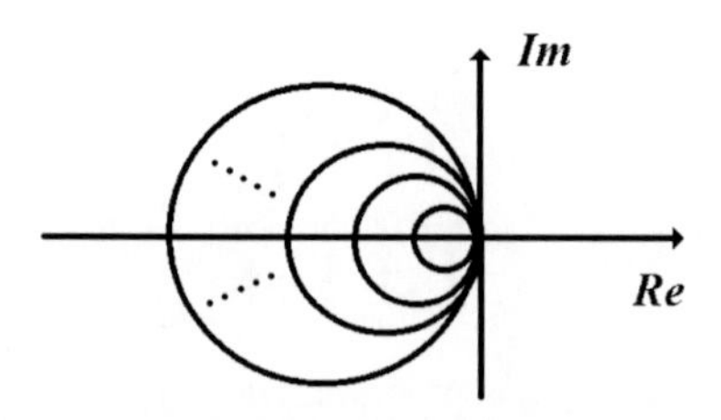

Figure 2. The Gershgorin circles $\bigcup_{i=1}^{n} \overline{D}(-s_i, s_i)$ corresponding to a FOCKM-matrix.

Moreover, all the Gershgorin circles are contained in the closure of the left semi-plane, $\overline{C}_{(-)}$, i.e.,

$$\bigcup_{i=1}^{n} \overline{D}(-s_i, s_i) \subset \{z \in C / \mathrm{Re}(z) \leq 0\} = \overline{C}_{(-)} \tag{17}$$

This situation is shown in Figure 2.

As a consequence of (33), the spectrum of A is also included in $\overline{C}_{(-)}$, or in other words:

$$\lambda_i \in \overline{C}_{(-)} \quad \forall i = 1,2,\ldots,n \tag{18}$$

According to the previous results, it follows that:

$$\mathrm{Re}(\lambda_i) \leq 0 \ \ \forall i = 1,2,\ldots,n \tag{19}$$

and

$$\mathrm{Re}(\lambda_k) = 0 \ \Leftrightarrow \ \lambda_k = 0 \tag{20}$$

This result is stated in the following proposition.

Proposition 2. If A is a $n \times n$ FOCKM-matrix and $\lambda_i \quad i = 1,2,\ldots,n$ are its eigenvalues, then $\mathrm{Re}(\lambda_i) \leq 0 \ \forall i = 1,2,\ldots,n$ and $\mathrm{Re}(\lambda_k) = 0 \Leftrightarrow \lambda_k = 0$.

From the ODE solutions point of view, there are three different cases to be considered, which have important implications for the qualitative behavior of the ODE system solutions.

Case 1. λ is a non-zero real eigenvalue (i.e., $\lambda < 0$)

In this case, the fundamental solutions are: $\exp(\lambda t), t\exp(\lambda t), t^2\exp(\lambda t), \ldots, t^p\exp(\lambda t)$, where p depends on the algebraic multiplicity (A.M.) and the geometric multiplicity (G.M.) of the eigenvalue λ. In this case $\lambda < 0$, so all these functions tend to vanish when $t \to \infty$, independently of p and the A.M. and G.M. corresponding to λ.

Case 2. $\lambda = a + ib$ is a non-zero complex eigenvalue (i.e, $a < 0$)

In this second case the fundamental solutions are of the ODE system are:

$$e^{at}\cos(bt), e^{at}\sin(bt), te^{at}\cos(bt), te^{at}\sin(bt), t^2e^{at}\cos(bt),$$
$$t^2e^{at}\sin(bt), \ldots, t^pe^{at}\cos(bt), t^pe^{at}\cos(bt)e^{at}$$

where, once again, p depends on the A.M. and G.M. of the eigenvalue λ. Considering that $a < 0$, all these functions tend to zero when $t \to \infty$, independently of p.

Case 3. $\lambda = 0$

The null eigenvalue is always present in FOCKM-problems as it was proved above. The ODE system solutions associated with the null eigenvalue are linear combinations of the following functions: $\{e^{0t}, te^{0t}, \ldots, t^qe^{0t}\}$, or the equivalent: $\{1, t, \ldots, t^q\}$. Then, the solutions due to the null eigenvalue are polynomial functions, and the corresponding polynomial degree q depends on both the A.M. and the G.M. associated to $\lambda = 0$.

It follows straightforwardly that if and only if $q = 0$, the polynomial solutions remain bounded when t tends to infinity.

Observation 1. To sum up the previous results, it can be stated that for a FOCKM-problem only the null eigenvalue – and particularly, its A.M. and G.M. – is relevant to make predictions about the stability of the ODE system solutions.

This problem concerning the $A.M._{\lambda=0}$ and the $G.M._{\lambda=0}$ is considered in the following proposition:

Proposition 3. In any FOCKM involving n chemical species, it must be $G.M._{\lambda=0} = A.M._{\lambda=0}$.

Proof. As mentioned above, the ODE system that corresponds to a general FOCKM is:

$$\begin{pmatrix} [E_1] \\ [E_2] \\ \vdots \\ [E_n] \end{pmatrix}' = \begin{pmatrix} -s_1 & k_{21} \cdots & k_{n1} \\ k_{12} & -s_2 \cdots & k_{n2} \\ \vdots & \vdots & \vdots \\ k_{1n} & k_{2n} \cdots & -s_n \end{pmatrix} \begin{pmatrix} [E_1] \\ [E_2] \\ \vdots \\ [E_n] \end{pmatrix} \tag{9}$$

If all these equations are added, it follows that $\frac{d}{dt}([E_1]+[E_2]+\ldots+[E_n])=0$ (which is a consequence of the equality $row_1 + row_2 + \ldots + row_n = \vec{0}$, already observed) and so, the sum of variables $[E_1]+[E_2]+\ldots+[E_n]$ is a constant κ.

Consequently, it follows that for every $[E_i]$ with $1 \le i \le n$, the inequalities $0 \le [E_i] \le [E_1]+[E_2]+\ldots+[E_n] = \kappa$ are satisfied and all $[E_i]$ must be bounded for every time $t \ge 0$.

If the null eigenvalue has different A.M. and G.M. – and so, the inequality $G.M._{\lambda=0} < A.M._{\lambda=0}$ takes place – then solutions of the ODE system will contain linear combinations of the following functions: $\{e^{0t}, te^{0t}, \ldots, t^q e^{0t}\}$, or the equivalent: $\{1, t, \ldots, t^q\}$ which are unbounded for $t \ge 0$. This fact obviously contradicts the previous statement about $[E_i]$ and consequently, it must be $G.M._{\lambda=0} = A.M._{\lambda=0}$ and the proposition is proved.

Observation 2. This result – valid for every FOCKM involving n chemical species – can be combined with the previous Observation 1 (i.e., only the null eigenvalue is relevant for stability issues) in order to get another important general result: the solutions of the FOCKM-problem ODE system are stable, but not asymptotically stable.

Observation 3. This weak stability observed in every FOCKM involving n chemical species, has an important chemical consequence, since it implies that small errors in the initial concentration measurements will remain bounded as time progresses, but they will not tend to disappear when $t \to +\infty$.

4. Characterization of MP-Matrices

A different kind of chemical problem considers mixtures, tanks, fluxes, compartments, etc., and it can be modeled by using linear ODE systems. The corresponding associated matrices will be called MP-matrices and their particular structure will be analyzed here. For this purpose we begin by analyzing a problem already considered in a previous paper [28] and in a recent Nova book chapter [7], devoted to the Laplace transform. The problem involves a system of three tanks (or compartments) shown in Figure 3.

Thus, C_0 is the initial concentration (for example, salt concentration in water at the entrance of the tank system), C_i is the concentration in the i-th compartment (i=1, 2, 3) and $\Phi_0 \neq 0$ is the incoming flux. This problem can be modeled by the following ODE system:

$$\begin{cases} V_1 \dfrac{dC_1}{dt} = \dfrac{1}{2}\Phi_0 C_0 - \Phi_1 C_1 \\ V_2 \dfrac{dC_2}{dt} = \dfrac{1}{2}\Phi_0 C_0 - \Phi_2 C_2 \\ V_3 \dfrac{dC_3}{dt} = \Phi_1 C_1 + \Phi_2 C_2 - \Phi_3 C_3 \end{cases} \tag{21}$$

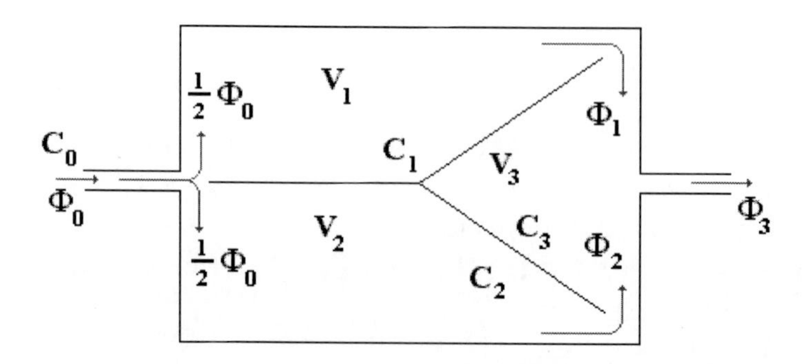

Figure 3. System of three tanks.

Thus, taking into account that Φ_0 must be equal to Φ_n since the compartments are neither filled up nor emptied with time, then, the corresponding mathematical model can be written as $\frac{d}{dt}\mathbf{C} = \mathbf{AC} + C_0\mathbf{B}$, where:

$$\mathbf{C} = \begin{pmatrix} C_1 \\ C_2 \\ C_3 \end{pmatrix}, \mathbf{A} = \begin{pmatrix} -\Phi_1/V_1 & 0 & 0 \\ 0 & -\Phi_2/V_2 & 0 \\ \Phi_1/V_3 & \Phi_2/V_3 & -\Phi_0/V_3 \end{pmatrix}$$

and

$$\mathbf{B} = \begin{pmatrix} \Phi_1/V_1 \\ \Phi_2/V_2 \\ 0 \end{pmatrix} \tag{22}$$

A first observation is that the MP-matrix associated with the ODE system, is an upper one with negative eigenvalues, so a first question is about the possibility of generalizing this particular result to other MP. Before trying to get general results, let us analyze a different problem, where a couple of tanks are linked by all the possible connections between them, including recirculation from the second tank back to the first one, as in Figure 4.

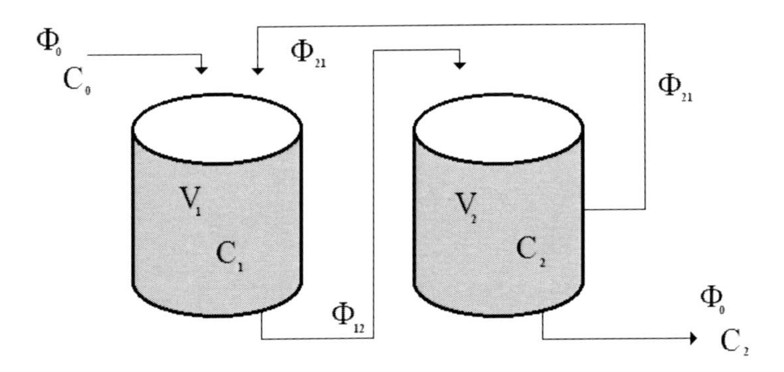

Figure 4. Two tanks with all the possible connections.

In this case, the ODE system can be formulated as:

$$\begin{cases} V_1 \dfrac{dC_1}{dt} = \Phi_0 C_0 + \Phi_{21} C_2 - \Phi_{12} C_1 \\ V_2 \dfrac{dC_2}{dt} = \Phi_{12} C_1 - \Phi_{21} C_2 - \Phi_0 C_2 \end{cases} \tag{23}$$

This mathematical model can be written as $\frac{d}{dt}\mathbf{C} = \mathbf{AC} + C_0\mathbf{B}$, where:

$$\mathbf{C} = \begin{pmatrix} C_1 \\ C_2 \end{pmatrix},\ \mathbf{A} = \begin{pmatrix} -\Phi_{12}/V_1 & \Phi_{21}/V_1 \\ \Phi_{12}/V_2 & -(\Phi_{21} + \Phi_0)/V_2 \end{pmatrix}$$

and

$$\mathbf{B} = \begin{pmatrix} \Phi_0 C_0 / V_1 \\ 0 \end{pmatrix} \tag{24}$$

Here, the MP-matrix is not an upper one and the signs of the eigenvalues are not obvious as in the previous case. Because of that, the general structure of an MP-matrix is more difficult to be described than FOCKM-matrices.

In order to give some general results, it is convenient to consider two different situations: MP without recirculation and MP with recirculation. Those cases will be analyzed separately in the following two sub-sections.

4.1. MP without Recirculation

Considering a tank with n compartments, like the example in Figure 5 (in this case $n = 5$). Let us suppose that it is possible to enumerate these compartments such that the flux always goes from the i-th compartment to the j-th one, where $i < j$. As an example, in Figure 5, a possible enumeration for this purpose could be the following: 1) left compartment, 2) upper compartment, 3) central compartment, i.e., the round one, 4) lower compartment, and 5) right compartment (in Figure 5, the given numbers are in brackets).

For instance, let us think about three particular examples: the ODE corresponding to the left compartment can be written as:

$$V_1 \frac{dC_1}{dt} = \Phi_0 C_0 - \Phi_{12} C_1 - \Phi_{13} C_1 - \Phi_{14} C_1 = \Phi_0 C_0 - (\Phi_{12} + \Phi_{13} + \Phi_{14}) C_1 \quad (25)$$

the ODE associated with the central compartment is:

$$V_3 \frac{dC_3}{dt} = \Phi_{13} C_1 - \Phi_{35} C_3 \quad (26)$$

and finally, for the right compartment we have:

$$V_5 \frac{dC_5}{dt} = \Phi_{25} C_2 + \Phi_{35} C_3 + \Phi_{45} C_4 - \Phi_0 C_5 \quad (27)$$

It is easy to observe that in all cases the ODE on the right hand side for the j-th compartment is a linear combination of a subset of $\{C_0, C_1, \ldots, C_{j-1}, C_j\}$ and this result can be extended straightforwardly as follows:

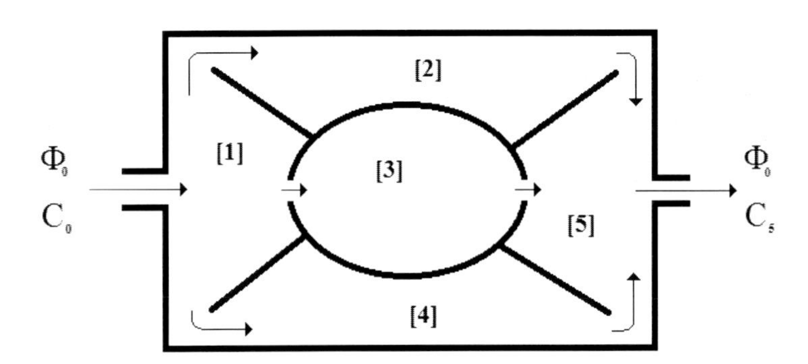

Figure 5. A tank with several internal compartments.

Proposition 4. If in a given MP the compartments can be enumerated such that there is no recirculation (i.e., if $i < j$ there is no flux from compartment j to compartment i), then the ODE corresponding to the j-th compartment will be of the form:

$$V_j \frac{dC_j}{dt} = \alpha_{i1} C_{i1} + \alpha_{i2} C_{i2} + \ldots + \alpha_{ik} C_{ik}$$

being

$$\{i1, i2, \ldots, ik\} \subset \{1, 2, \ldots, j\} \text{and } \alpha_{i1}, \alpha_{i2}, \ldots, \alpha_{ik} \in R.$$

As a consequence, the following corollary can be stated:

Corollary 2. In the conditions of the previous proposition, the corresponding ODE system has an associated upper matrix.

Furthermore, equations (25), (26) and (27) can be re-written as:

$$\frac{dC_1}{dt} = \frac{\Phi_0}{V_1} C_0 - \frac{\Phi_{12} + \Phi_{13} + \Phi_{14}}{V_1} C_1 \tag{28}$$

$$\frac{dC_3}{dt} = \frac{\Phi_{13}}{V_3} C_1 - \frac{\Phi_{35}}{V_3} C_3 \tag{29}$$

and

$$\frac{dC_5}{dt} = \frac{\Phi_{25}}{V_5} C_2 + \frac{\Phi_{35}}{V_5} C_3 + \frac{\Phi_{45}}{V_5} C_4 - \frac{\Phi_0}{V_5} C_5 \tag{30}$$

It follows that for the j-th compartment, the coefficient corresponding to C_j can be written as $\frac{-\sum_k \Phi_k}{V_j}$, where $\sum_k \Phi_k$ represents the sum of fluxes leaving the tank compartment. This situation can be easily generalized, since concentration C_j only appears in the right hand side of the corresponding ODE when a certain flux is leaving the tank. Combining this result with the previous corollary 2, a new corollary can be stated as follows:

Corollary 3. In the conditions of the previous proposition, the ODE system has only negative eigenvalues of the form $\lambda_j = \frac{-\sum_k \Phi_k}{V_j} < 0$,for all $j = 1, 2, \ldots, n$.

Not all of these results can be extended to MP with recirculation as will be analyzed in the following section.

4.2. MP with Recirculation

As mentioned above, the ODE system corresponding to an MP involving two tanks with recirculation (i.e., $\Phi_{21} \neq 0$) can be written as: $\frac{d}{dt}\mathbf{C} = \mathbf{AC} + C_0\mathbf{B}$, where:

$$\mathbf{C} = \begin{pmatrix} C_1 \\ C_2 \end{pmatrix}, \ \mathbf{A} = \begin{pmatrix} -\Phi_{12}/V_1 & \Phi_{21}/V_1 \\ \Phi_{12}/V_2 & -(\Phi_{21} + \Phi_0)/V_2 \end{pmatrix}$$

and

$$\mathbf{B} = \begin{pmatrix} \Phi_0 C_0/V_1 \\ 0 \end{pmatrix} \tag{24}$$

In this situation, the associated matrix is not an upper one, because there is recirculation, $\Phi_{21} \neq 0$, and then the matrix entry $a_{21} = \frac{\Phi_{21}}{V_1}$ is not null, so the previous proposition and corollaries do not hold.

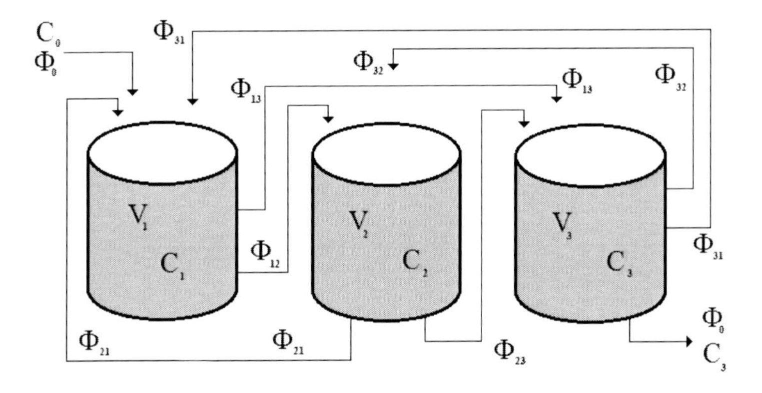

Figure 6. Three tanks with all the possible connections.

In order to know how recirculation modifies the previous conclusions, let us examine a three tanks system with all the possible connections among them, as in Figure 6.

In this case, the ODE system can be written as follows:

$$\begin{cases} V_1 \dfrac{dC_1}{dt} = \Phi_0 C_0 + \Phi_{21} C_2 + \Phi_{31} C_3 - (\Phi_{12} + \Phi_{13}) C_1 \\ V_2 \dfrac{dC_2}{dt} = \Phi_{12} C_1 + \Phi_{32} C_3 - (\Phi_{21} + \Phi_{23}) C_2 \\ V_3 \dfrac{dC_3}{dt} = \Phi_{13} C_1 + \Phi_{23} C_2 - (\Phi_{31} + \Phi_{32} + \Phi_0) C_3 \end{cases} \tag{31}$$

The corresponding mathematical model can be expressed once again, in the form $\frac{d}{dt}\mathbf{C} = \mathbf{AC} + C_0 \mathbf{B}$, where in this case:

$$\mathbf{C} = \begin{pmatrix} C_1 \\ C_2 \\ C_3 \end{pmatrix},$$

$$\mathbf{A} = \begin{pmatrix} (-1/V_1)(\Phi_{12} + \Phi_{13}) & \Phi_{21}/V_1 & \Phi_{31}/V_1 \\ \Phi_{12}/V_2 & (-1/V_2)(\Phi_{21} + \Phi_{23}) & \Phi_{32}/V_2 \\ \Phi_{13}/V_3 & \Phi_{23}/V_3 & (-1/V_3)(\Phi_{31} + \Phi_{32} + \Phi_0) \end{pmatrix}$$

and

$$\mathbf{B} = \begin{pmatrix} \Phi_0 / V_1 \\ 0 \\ 0 \end{pmatrix} \tag{32}$$

It is easy to observe that the MP-matrix $\mathbf{A}$ is not an upper one, if any of the recirculation fluxes $\Phi_{21}, \Phi_{23}, \Phi_{32}$ is not zero.

As a consequence, if the MP includes recirculation, the structure of the MP-matrix is not an upper matrix. However, it is possible to give a necessary condition to be satisfied by any MP-matrix. For this purpose, let us analyze a "black box" (Figure 7), where the number of compartments and the internal geometry are unknown:

Here Φ_0 and C_0 represent flux and concentration at the input, Φ_0 is also the output flux (tanks neither fill up nor empty with time) and C_n is the final concentration. In this system there are n compartments in the black box with volumes V_i and concentrations C_i and recirculation fluxes may exist or not. Hence, if m represents the mass in the black box, then:

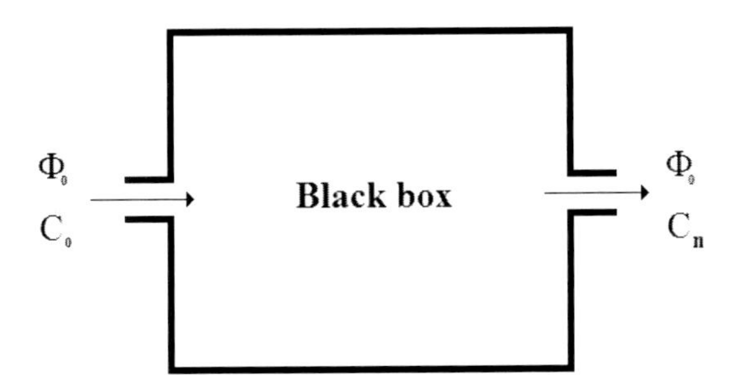

Figure 7. A "black box" tank system.

$$m = \sum_{i=1}^{n} V_i C_i \tag{33}$$

If all volumes V_i remain constant a mass balance gives:

$$\frac{dm}{dt} = \Phi_0 C_0 - \Phi_0 C_n = \Phi_0 (C_0 - C_n) \tag{34}$$

On the other hand, by differentiation of (34), the following equation can be obtained:

$$\frac{dm}{dt} = \sum_{i=1}^{n} V_i \frac{dC_i}{dt} \tag{35}$$

Finally, combining equations (34) and (35), we obtain:

$$\sum_{i=1}^{n} V_i \frac{dC_i}{dt} = \Phi_0 (C_0 - C_n) \tag{36}$$

Equation (36) was obtained without any consideration of the internal geometry of the tanks system and can be easily verified by adding equations of the ODE systems (21), (23) and (31), among other possible examples. Hence, this result can be stated as the following proposition:

Proposition 5. In a given MP – with or without recirculation – with input and output concentrations C_0 and C_n, respectively, and where $\Phi_0 \neq 0$ is the incoming and outgoing flux then, independently of the internal geometry the equation $\sum_{i=1}^{n} V_i \frac{dC_i}{dt} = \Phi_0 (C_0 - C_n)$ is satisfied.

It is interesting to note that if $\mathbf{V} = (V_1, V_2, \ldots, V_n)$ is the volume's vector, then equation (36) can also be written as:

$$\mathbf{V}^{\mathrm{T}} \frac{d\mathbf{C}}{dt} = \sum_{i=1}^{3} V_i \frac{dC_i}{dt} = \Phi_0 (C_0 - C_n) \tag{37}$$

Equation (37) is a condensed version of these two equalities:

$$\mathbf{V}^{\mathrm{T}} \mathbf{AC} = -\Phi_0 C_n \tag{38}$$

and

$$C_0 \mathbf{V}^{\mathrm{T}} \mathbf{B} = \Phi_0 C_0 \tag{39}$$

Moreover, it is very easy to show that

$$\mathbf{V}^{\mathrm{T}} \mathbf{A} = \begin{pmatrix} 0 & 0 & -\Phi_0 \end{pmatrix} \tag{40}$$

and

$$\mathbf{V}^{\mathrm{T}} \mathbf{B} = \Phi_0 \tag{41}$$

which only depend on the incoming and outgoing flux. Equation (40) is particularly interesting, since it can be used to characterize MP-matrices as mentioned in the following proposition.

Proposition 6. In a given MP – with or without recirculation – where Φ_0 is the incoming and outgoing flux, the associated MP-matrix satisfies the equality $\mathbf{V}^{\mathrm{T}}\mathbf{A} = \begin{pmatrix} 0 & 0 & -\Phi_0 \end{pmatrix}$, independently of the internal geometry of the system.

As previously mentioned, this proposition may be used to know if a given matrix can or cannot be an MP-matrix corresponding to a certain tanks system. For example, let us consider the MP-matrix $\mathbf{A}$, already introduced in (22):

$$\mathbf{A} = \begin{pmatrix} -\Phi_1/V_1 & 0 & 0 \\ 0 & -\Phi_2/V_2 & 0 \\ \Phi_1/V_3 & \Phi_2/V_3 & -\Phi_0/V_3 \end{pmatrix} \tag{22}$$

It is easy to show that $\mathbf{A}$ satisfies Proposition 6, i.e., $\mathbf{V}^{\mathrm{T}}\mathbf{A} = \begin{pmatrix} 0 & 0 & -\Phi_0 \end{pmatrix}$. However, if $\mathbf{A}$ is slightly changed – for example, in its first entry – the new matrix will not satisfy that condition. To see this result, let us consider the matrix $\mathbf{A}_\varepsilon$:

$$\mathbf{A}_\varepsilon = \begin{pmatrix} \left(-\Phi_1/V_1\right)+\varepsilon & 0 & 0 \\ 0 & -\Phi_2/V_2 & 0 \\ \Phi_1/V_3 & \Phi_2/V_3 & -\Phi_0/V_3 \end{pmatrix},$$

which gives $\mathbf{V}^{\mathrm{T}}\mathbf{A}_\varepsilon = \begin{pmatrix} \varepsilon V_1 & 0 & -\Phi_0 \end{pmatrix}$.

It is easily observed that the modified matrix does not verify Proposition 6, so there is no MP associated to this matrix $\mathbf{A}_\varepsilon$. In consequence, not every matrix is an MP-matrix and slight changes in a given MP-matrix, produces a non-MP-matrix. Then, existence and stability questions for the inverse-modeling problem have negative answers.

Furthermore, if volumes V_i and fluxes Φ_i are multiplied by a scale factor, then the MP-matrix $\mathbf{A}$ remains unchanged. So, a scale factor in geometry, not in concentrations, produces exactly the same mathematical model, giving a new negative answer to the question of uniqueness.

Besides, as in chemical kinetics problems, existence, uniqueness and stability do not occur when working with MP-matrices [12, 24].

5. Stability Properties of MP-Matrices ODE Systems

As stated in Corollary 3, if there is no recirculation, then the ODE system has only negative eigenvalues of the form $\lambda_j = \frac{-\sum_k \Phi_k}{V_j} < 0$, for all $j = 1, 2, \ldots, n$.

Then, in this case the following proposition is obtained straightforwardly:

Proposition 7. If in a given MP the compartments can be enumerated such that there is no recirculation (i.e., if $i < j$ there is no flux from compartment j to compartment i), then all the corresponding ODE system solutions will be asymptotically stable.

It is important to know if Proposition 7 is also true for MP with recirculation. In order to investigate this possibility, it is useful to go back firstly to the MP-matrix

$$\mathbf{A} = \begin{pmatrix} -\Phi_{12}/V_1 & \Phi_{21}/V_1 \\ \Phi_{12}/V_2 & -(\Phi_{21}+\Phi_0)/V_2 \end{pmatrix} \tag{24}$$

corresponding to a couple of tanks with recirculation.

For this MP-matrix $\mathbf{A}$, we have that $tr(\mathbf{A}) = -\Phi_{12}/V_1 - (\Phi_{21}+\Phi_0)/V_2 < 0$ and

$$\det(\mathbf{A}) = [-\Phi_{12}/V_1][-(\Phi_{21}+\Phi_0)/V_2] - [\Phi_{21}/V_1][\Phi_{12}/V_2] = \frac{\Phi_{12}.\Phi_0}{V_2 V_2} > 0.$$

These inequalities are always satisfied, except if $\Phi_{12} = 0$ and/or $\Phi_0 = 0$ .In the first case ($\Phi_{12} = 0$) there is no flux from the first tank to the second, while in the other case ($\Phi_0 = 0$) there is no incoming flux. Obviously, the tank system does not work or there is no incoming flux, and then, in all non trivial cases $tr(\mathbf{A}) < 0$ and $\det(\mathbf{A}) > 0$.

In addition to this, a well known property states that if a 2×2 matrix $\mathbf{A}$ has two eigenvalues λ_1 , λ_2 , then, the trace of $\mathbf{A}$ is $\lambda_1 + \lambda_2$ and its determinant is $\lambda_1 . \lambda_2$. Consequently, if the eigenvalues are real numbers, the inequality $\det(\mathbf{A}) > 0$ implies that λ_1 and λ_2 must be both positive or both negative, and so, the other inequality $tr(\mathbf{A}) < 0$ implies that $\lambda_1 < 0$, $\lambda_2 < 0$. Lastly, if the eigenvalues are complex numbers $\lambda_1 = a + ib$ and $\lambda_2 = a - ib$, then the inequality about $tr(\mathbf{A})$ implies that $tr(\mathbf{A}) = 2a < 0 \Rightarrow a < 0$ and so $\mathrm{Re}(\lambda_1) < 0$, $\mathrm{Re}(\lambda_2) < 0$. As a consequence, in all cases the ODE system solutions will be asymptotically stable.

Also, another case that deserves to be analyzed corresponds to the problem with three tanks with all the possible connections among them. The associated MP-matrix is:

$$\mathbf{A} = \begin{pmatrix} (-1/V_1)(\Phi_{12} + \Phi_{13}) & \Phi_{21}/V_1 & \Phi_{31}/V_1 \\ \Phi_{12}/V_2 & (-1/V_2)(\Phi_{21} + \Phi_{23}) & \Phi_{32}/V_2 \\ \Phi_{13}/V_3 & \Phi_{23}/V_3 & (-1/V_3)(\Phi_{31} + \Phi_{32} + \Phi_0) \end{pmatrix} \quad (32)$$

If the Gershgorin circle theorem is applied to this matrix, we will obtain three closed discs of the form $\mathbf{D}(a_{ii}, R_i)$ where the center a_{ii} is the i-th diagonal entry and the radius R_i can be computed as $R_i = \sum_{j \neq i} |a_{ij}|$.

The first disc $\mathbf{D}_1$ is centered on $a_{11} = -\dfrac{\Phi_{12} + \Phi_{13}}{V_1}$ with a radius $R_1 = \dfrac{\Phi_{21} + \Phi_{31}}{V_1}$. A flux balance in the first tank of Figure 6 gives

$$\Phi_{12} + \Phi_{13} = \Phi_0 + \Phi_{21} + \Phi_{31} \quad (42)$$

and then $\Phi_{21} + \Phi_{31} < \Phi_{12} + \Phi_{13}$, and so $R_1 < |a_{11}|$. The second disc $\mathbf{D}_2$ is centered on $a_{22} = -\dfrac{\Phi_{21} + \Phi_{23}}{V_2}$ with a radius $R_2 = \dfrac{\Phi_{12} + \Phi_{32}}{V_2}$ and again a flux balance in the second tank of Figure 6 gives

$$\Phi_{12}+\Phi_{32}=\Phi_{21}+\Phi_{23} \tag{43}$$

and so $R_2=|a_{22}|$. Finally, the third disc $\mathbf{D}_3$ is centered on $a_{33}=-\dfrac{\Phi_{31}+\Phi_{32}+\Phi_0}{V_3}$ with a radius $R_3=\dfrac{\Phi_{13}+\Phi_{23}}{V_3}$ and the corresponding flux balance in the third tank of Figure 6 gives

$$\Phi_{13}+\Phi_{23}=\Phi_{31}+\Phi_{32}+\Phi_0 \tag{44}$$

and so $R_3=|a_{33}|$. Taking into account all these results, the Gershgorin circles for this MP can be schematized as in Figure 8.

Since every eigenvalue lies within at least one of the Gershgorin discs, it follows that $\mathrm{Re}(\lambda_i)\leq 0$, $\forall i=1,2,3$.

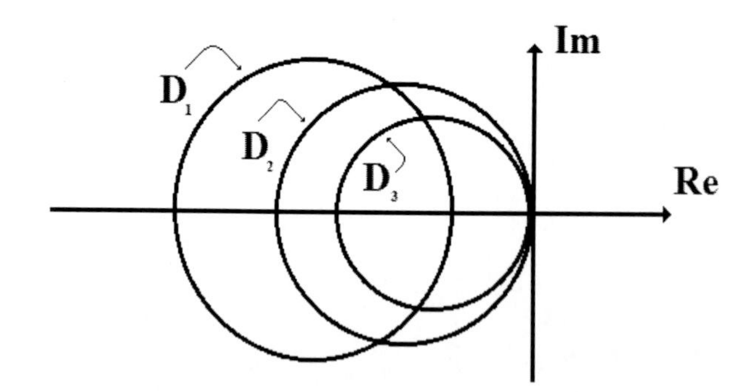

Figure 8. Gershgorin circles for a three tank system with recirculation.

It is well known that $\lambda=0$ is an eigenvalue of the matrix $\mathbf{A}$ if and only if $\det(\mathbf{A})=0$, so if we prove that $\det(\mathbf{A})\neq 0$, then all the eigenvalues will have negative real part, as occurs in the two tanks problem.

The MP-matrix determinant is:

$$\det(\mathbf{A})=\begin{vmatrix} (-1/V_1)(\Phi_{12}+\Phi_{13}) & \Phi_{21}/V_1 & \Phi_{31}/V_1 \\ \Phi_{12}/V_2 & (-1/V_2)(\Phi_{21}+\Phi_{23}) & \Phi_{32}/V_2 \\ \Phi_{13}/V_3 & \Phi_{23}/V_3 & (-1/V_3)(\Phi_{31}+\Phi_{32}+\Phi_0) \end{vmatrix} \tag{45}$$

If the first and second columns are added to the third one – and the flux balances (42), (43) and (44) are taken into account – we obtain that:

$$\det(\mathbf{A}) = \begin{vmatrix} (-1/V_1)(\Phi_{12}+\Phi_{13}) & \Phi_{21}/V_1 & -\Phi_0/V_1 \\ \Phi_{12}/V_2 & (-1/V_2)(\Phi_{21}+\Phi_{23}) & 0 \\ \Phi_{13}/V_3 & \Phi_{23}/V_3 & 0 \end{vmatrix} \quad (46)$$

Which gives the following:

$$\det(\mathbf{A}) = -\left(\frac{\Phi_0}{V_1}\right)\begin{vmatrix} \Phi_{12}/V_2 & (-1/V_2)(\Phi_{21}+\Phi_{23}) \\ \Phi_{13}/V_3 & \Phi_{23}/V_3 \end{vmatrix} \quad (47)$$

After performing some algebraic manipulations, the final result is:

$$\det(\mathbf{A}) = -\left(\frac{\Phi_0}{V_1}\right)\frac{\Phi_{12}\Phi_{23}+\Phi_{13}\Phi_{21}+\Phi_{13}\Phi_{23}}{V_2V_3} \quad (48)$$

and $\det(\mathbf{A})=0$ if and only if $\Phi_0=0$ and/or $\Phi_{12}\Phi_{23}+\Phi_{13}\Phi_{21}+\Phi_{13}\Phi_{23}=0$. In the first situation there will be no incoming flux and this is not an MP of our interest. In the second one, we can consider three sub-cases, like in the tree diagram in Figure 9.

If there is some incoming flux (i.e., $\Phi_0 \neq 0$), then $\det(\mathbf{A})=0$ if and only if $\Phi_{12}\Phi_{23}+\Phi_{13}\Phi_{21}+\Phi_{13}\Phi_{23}=0$. The case 1 – in the previous tree – is a possible form of getting this result, since $\Phi_{12}=\Phi_{13}=0$. If the flux balance $\Phi_{12}+\Phi_{13}=\Phi_0+\Phi_{21}+\Phi_{31}$ (42) is taken into account, this situation only takes place if there is no incoming flux. Another option is given by case 2 (see Figure 9), where $\Phi_{13}=\Phi_{23}=0$, and so, if the flux balance $\Phi_{13}+\Phi_{23}=\Phi_{31}+\Phi_{32}+\Phi_0$ (44) is considered, this fact does not take place if there is a non null incoming flux. In the third case of the previous tree we have $\Phi_{13}\neq 0$, so the equation $\Phi_{12}\Phi_{23}+\Phi_{13}\Phi_{21}+\Phi_{13}\Phi_{23}=0$ only can be satisfied if $\Phi_{21}=\Phi_{23}=0$. Then, if the flux balance $\Phi_{12}+\Phi_{32}=\Phi_{21}+\Phi_{23}$ (43) is taken into account, it must be also that $\Phi_{12}+\Phi_{32}=0$ and finally $\Phi_{12}=\Phi_{32}=0$. In this hypothetical situation

$\Phi_{12} = \Phi_{32} = \Phi_{21} = \Phi_{23} = 0$ and the second tank would be isolated from the rest of the components of the system.

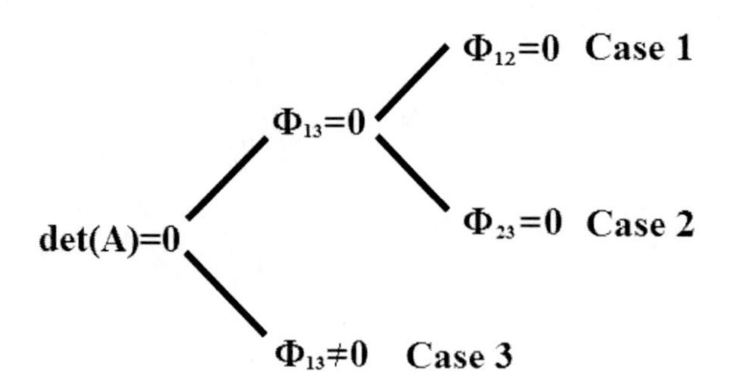

Figure 9. Tree diagram for analyzing if $\det(\mathbf{A}) = 0$.

Therefore, from the previous analysis, it is proved that $\det(\mathbf{A}) \neq 0$ in all the non-trivial cases and then all the eigenvalues will have negative real part, and then, all the ODE system solutions are asymptotically stable.

All these results can be included in the following statement:

Proposition 8. In a given MP with three or less compartments with or without recirculation, all the corresponding ODE system solutions will be asymptotically stable.

It is important to mention that the case with only one compartment is also included in this proposition, since in this situation there is no recirculation, and it represents a particular case of Proposition 7.

Conclusion

Chemical kinetics and mixing problems are interesting sources for applied research in mathematical modeling, ODE and linear algebra. In FOCKM-problems the sum of all the differential equations becomes zero, whereas in MP, a linear combination of all the system equations gives $\Phi_0(C_0 - C_n)$, i.e., the right side of equation (36). The last expression only vanishes in trivial cases: $\Phi_0 = 0$ (no incoming flux) and/or $C_0 = C_n$ (no changes in concentration after leaving the system).

Another important difference appears when FOCKM and MP matrices are studied, since $\lambda = 0$ is an eigenvalue of any FOCKM-matrix $\mathbf{A}$, however null eigenvalues are not expected in MP-matrices, at least in problems with three or less components.

The Gershgorin Circles Theorem can be applied to both FOCKM and MP matrices. In the first case a general conclusion – $\mathrm{Re}(\lambda_i) \leq 0$, $\forall i$ – can be obtained, while in the second case, the same result was only proved if there is no recirculation and/or three or less compartments are considered.

An observation worth noting is that chemical kinetics problems are more interesting from the non-linearity point of view, since only FOCKM-problems give linear ODE systems. However, MP show more mathematical richness when linear ODE systems are considered. In particular, an interesting challenge consists in extending the previous results to MP with more than three tanks, with recirculation among them.

Regarding the inverse modeling problem, FOCKM and MP showed similar behavior. Firstly, given an arbitrary ODE system, it is impossible to assure that there exists an associated FOCKM or MP. Secondly, if the corresponding FOCKM or MP exists, it may not be unique and finally, if the ODE system is slightly changed then there will be no FOCKM nor MP associated. So, in both cases existence, uniqueness and stability have negative results.

As a final remark, the qualitative behavior of FOCKM and MP differential equations systems solutions are different. In FOCKM-problems these solutions are stable, but not asymptotically stable, while in MP the ODE solutions are asymptotically stable, at least in all the cases studied in this chapter.

Acknowledgments

The author wishes to thank Marjorie Chaves and Valerie Dee for their assistance and support in this work.

References

[1] Mickley, H.S., Sherwood, T.S. and Reed, C.E. (1975). *Applied Mathematics in Chemical Engineering, 2º Ed.*, New Delhi: Tata Mc Graw-Hill Publ. Co. Ltd.

[2] Westerterp, K.R., (1984). *Chemical Reactor Design and Operations*. New Jersey: Wiley.

[3] Martinez-Luaces, V., (2015). First Order Chemical Kinetics Matrices and Stability of O.D.E. Systems, in: *Advances in Linear Algebra Research*. Chapter 10, pp. 325-343. New York: Nova Publishers.

[4] Martinez-Luaces, V., (2016). Inverse Modeling Problems and their Potential in Mathematics Education, in: *Teaching and Learning: Principles, Approaches and Impact Assessment*, Chapter 7. New York: Nova Publishers.

[5] Martinez-Luaces, V., (2017). Qualitative Behavior of Concentration Curves in First Order Chemical Kinetics Mechanisms, in: *Advances in Chemistry Research, Volume 34*, Chapter 7. New York: Nova Publishers.

[6] Martinez-Luaces, V., (2017). Relevance, motivation and meaningful learning in Mathematics Education, in: *Progress in Education, Volume 45*, Chapter 7. New York: Nova Publishers.

[7] Martinez-Luaces, V., (2017). Laplace Transform in Chemistry degrees Mathematics courses in *Focus on Mathematics Educations Research*, Chapter 4. New York: Nova Publishers.

[8] Martinez-Luaces, V., (2009). Modelling, applications and Inverse Modelling: Innovations in Differential Equations courses, in *Proceedings of Delta '09, Seventh Southern Hemisphere Conference on Undergraduate Mathematics and Statistics Teaching and Learning.*South Africa: ISC-Delta.

[9] Martinez-Luaces, V., (2012). *Problemas inversos y de modelado inverso en Matemática Educativa* [inverse and inverse modeling problems in Mathematics Education]. Saarbrücken, Germany: Editorial Académica Española.

[10] Martinez-Luaces, V., (2003). Mass Transfer: the other half of Parabolic P.D.E. *New Zealand Journal of Mathematics*, *32*, 125-133.

[11] Martinez-Luaces, V., (2009). Modelling and inverse modelling: innovations with second order P.D.E. in engineering courses, in *Proceedings of Delta '09, Seventh Southern Hemisphere Conference on Undergraduate Mathematics and Statistics Teaching and Learning.* South Africa: ISC-Delta.

[12] Martinez-Luaces, V., (2007). Inverse-modelling problems in chemical engineering courses, in *Vision and change for a new century. Proceedings of the Sixth Southern Hemisphere Conference on*

Undergraduate Mathematics and Statistics Teaching and Learning. Montevideo, Uruguay: ISC-Delta.

[13] Martinez-Luaces, V., (2012). Chemical Kinetics and Inverse Modelling Problems, in *Chemical Kinetics*, Chapter 3, pp. 61-78.Rijeka, Croatia: In Tech Open Science.

[14] Ohanian, M. and Martinez-Luaces, V., (2014). Corrosion potential profile simulation in a tube under cathodic protection. *International Journal of Corrosion.* Article ID102363. http://dx.doi.org/10.1155/2014/102363.

[15] Ohanian, M., Martinez-Luaces, V. and Diaz, V., (2010). Trend removal from Electrochemical Noise data. *The Journal of Corrosion Science and Engineering.*, Vol. 13, Paper No. 52.

[16] Martinez-Luaces, M., Martinez-Luaces, V. and Ohanian, M., (2006). Trend-removal with Neural Networks: data simulation, preprocessing and different training algorithms applied to Electrochemical Noise studies. *WSEAS Transactions on Information Science and Applications,* Issue 4, Volume 3, pp. 810-817.

[17] Ohanian, M., Martinez-Luaces, V. and Guineo, G., (2004). Highly dispersed electrochemical noise data: searching for reasons and possible solutions. *The Journal of Corrosion Science and Engineering*, Vol. 7, Paper No. 11.

[18] Díaz V, Martinez-Luaces, V. and Guineo Cobs, G., (2003). Corrosión Atmosférica: validación de modelos empleando técnicas estadísticas [Atmospheric corrosion: validation of models using statistical techniques]. *Metalurgia,* 39.4, 243-251.

[19] Martinez-Luaces, V., Guineo Cobs, G., Velásquez, B., Chabalgoity, A. and Massaldi, H., (2006). Bifactorial Design applied to recombinant protein expressions. *Journal of Data Science.*Volume 4, Number 2, pp. 247-255.

[20] Velázquez, B., Martinez-Luaces, V., Vázquez, A., Dee, V. and Massaldi, H., (2007). Experimental Design techniques applied to study of oxigen consumption in a fermenter. *Journal of Applied Quantitative Methods.* Volume 2, Number 1, pp. 135-141.

[21] Martinez-Luaces, V., Velazquez, B. and Dee, V. (2009). A course on experimental design for different university specialties: experiences and changes over a decade, *International Journal of Mathematical Education in Science and Technology*, 40(5), 641-657.

[22] Martinez-Luaces, V., (2015). Stability of O.D.E. solutions corresponding to chemical mechanisms based-on unimolecular first order reactions, *Mathematical Sciences and Applications E-notes.* Vol. 3, No. 2, pp. 58-63.

[23] Martinez-Luaces, V., (2016). Stability of O.D.E. systems associated to first order chemical kinetics mechanisms with and without final products, *Konuralp Journal of Mathematics.* Vol. 4, No. 1, pp. 80-87.

[24] Martinez-Luaces, V., (2009). Modelling and inverse-modelling: experiences with O.D.E. linear systems in engineering courses. *International Journal of Mathematical Education in Science and Technology 40(2)*, 259-268.

[25] Courant, R. (1937). *Differential and Integral Calculus.* Volume I, 2nd Edition. London and Glasgow: Blackie & Son Limited.

[26] Wilhelmy, L. (1850), Über das Gesetz, nachwelchem die Einwirkung der Säuren auf den Rohrzuckerstattfindet, *Pogg. Ann.* Vol. 81, pp. 413-433. Available from http://gallica.bnf.fr/ark:/12148/bpt6k15166k/f427.table.

[27] Varga, R.S. (2004), *Geršgorin and His Circles.* Berlin, Germany: Springer-Verlag.

[28] Martinez-Luaces, V., (2005), Engaging Secondary School and University Teachers in Modelling: Some Experiences in South American Countries, *International Journal of Mathematical Education in Science and Technology*, *36(2-3),* 193-205.

Biographical Sketch

Victor Martinez-Luaces

Affiliation: Electrochemistry Multidisciplinary Research Group, Faculty of Engineering, UdelaR, Montevideo, Uruguay.

Education

Degree in Chemistry, Faculty of Chemistry, UdelaR, 1982.
Chemical Engineer, Faculty of Engineering, UdelaR, 1985.
Degree in Mathematics, Faculty of Sciences, UdelaR, 1992.
Post-graduate studies in Mathematics at IMPA (Brazil).

Post-graduate studies in Mathematics Education at UGR (Spain).

E-mail address: victorml@fing.edu.uy.

Research and Professional Experience

First book chapter, 1982, Uruguay.
First research paper, 1988, Journal of Chemical Education, U.S.A.
Since then, several papers have been published worldwide in the field of Mathematics Education, Chemistry, Engineering and Mathematics in journals, conference proceedings and books.
More than 15 books, 50 journal papers and 80 proceedings papers.

Professional Appointments

Faculty of Chemistry, UdelaRHead of the Mathematics Department 1996 – 2002.
Faculty of Engineering, UdelaR1988 – 1997.
Faculty of Economics, UdelaR1990 -2006.
Faculty of Sciences, UdelaR1999- 2000.
CIEP (Education & Research Institute) Rio Gallegos, Argentina2006-2008.
Since 2009 Mathematical and Engineering Consultant at J. Ricaldoni Foundation and the Electrochemistry Multidisciplinary Research Group, UdelaR.

Honors

- Journal of Chemical Education – (Ethyl Corporation, 1988).
- Best teacher award (Faculty of Engineering, 1992).
- Best teacher award (Faculty of Engineering, 1996).
- 2nd prize RELME 13 (Santo Domingo, Dominican Republic, 1999).
- 2ndprize RELME 14 (Panama, 2000).
- 2ndprize RELME 15 (Buenos Aires, Argentina, 2001).
- 1stprize RELME 16 (La Habana, Cuba, 2002).
- 1st prize RELME 19 (Montevideo, Uruguay, 2005).
- Distinguished visitor of the first city in the Americas (Santo Domingo, 1999).

- Convenor, Delta 07, Calafate, Argentina, 2007.
- Member of the Scientific Committee, Delta 09, South Africa, 2009.
- Member of the Scientific Committee, Delta 17, Gramado, Brazil, 2017.
- Co-chair of the TSG 13 at ICME-12, Seoul, Korea 2012.
- Co-chair of the TSG 16 at ICME-13, Hamburg, Germany 2016.
- Reviewer for several journals such as ZDM, iJMEST, IJRUME, ALME (Mexico), Relime (Mexico), Union (Spain) and Numeros (Spain), American Journal of Applied Mathematics and Statistics (U.S.A.), American Journal of Educational Research (U.S.A.), among others.
- Member of the Editorial Board for journals like IJRUME, Numeros, Union and Global Education Review.
- Guest Editor of iJMEST, October 2007.
- Guest Editor of iJMEST, July 2013.

Publications Last 3 Years

A. Books and Book Chapters

[1] Martinez-Luaces, V., (2017). Qualitative Behavior of Concentration Curves in First Order Chemical Kinetics Mechanisms in *Advances in Chemistry Research, Volume 34,* Chapter 7, New York: Nova Publishers.

[2] Martinez-Luaces, V., (2017). Relevance, motivation and meaningful learning in Mathematics Education in *Progress in Education, Volume 45,* Chapter 7, New York: Nova Publishers.

[3] Martinez-Luaces, V., (2017). Laplace Transform in Chemistry degrees Mathematics courses in *Focus on Mathematics Educations Research,* Chapter 4, New York: Nova Publishers.

[4] Martinez-Luaces, V., (2017) A Curriculum Design Decision as the Starting Point for a Multidisciplinary Research Group. In: *Success in Higher Education.* Singapore: Springer.

[5] Martinez-Luaces, V., (2016). Inverse Modeling Problems and Their Potential in Mathematics Education in *Teaching and Learning: Principles, Approaches and Impact Assessment*. Chapter 7, New York: Nova Publishers.

[6] Bressoud, D., Ghedamsi, I., Martinez-Luaces, V. & Törner, G., (2016). *Teaching and Learning of Calculus*. Berlin: Springer.
[7] Martinez-Luaces, V., (2016). *Problemas de modelado inverso en Educación Matemática* [Inverse modeling problems in Mathematics Education], Saarbrücken, Germany: Editorial Académica Española.
[8] Martinez-Luaces, V., (2015). First Order Chemical Kinetics Matrices and Stability of O.D.E. Systems in *Advances in Linear Algebra Research*. Chapter 10, pp. 325-343, New York: Nova Publishers.

B. Journal Publications

[1] Martinez-Luaces, V., (2016). Stability of ODE systems associated with first order chemical kinetics mechanisms with and without final products. *Konuralp Journal of Mathematics*, Vol. 4, No. 1, pp 80-87.
[2] Martinez-Luaces, V., (2015). Stability of ODE solutions corresponding to chemical mechanisms based-on unimolecular first order reactions. *Mathematical Sciences and Applications E-notes*, Vol. 3, No. 2, pp 58-63.
[3] Ohanian, M. & Martinez-Luaces, V., (2014). Corrosion potential profile simulation in a tube under cathodic protection. *International Journal of Corrosion*. Article ID102363. http://dx.doi.org/10.1155/2014/102363.

C. Proceedings and Other Publications

[1] Martinez-Luaces, V., (2015). Mathematical models for Chemistry and Biochemistry service courses. Proceedings of CERME 2015, pp 904-909. Prague, Czech Republic.
[2] Martinez-Luaces, V. & Noh, S., (2015). Report of the Topic Study Group 13 about Teaching and Learning of Calculus, Intellectual and attitudinal challenges, Proceedings of ICME 2012, pp. 447-452. Seoul, South Korea.

In: Advances in Mathematics Research
Editor: Albert R. Baswell

ISBN: 978-1-53612-512-2

Chapter 2

DETERMINANTAL REPRESENTATIONS OF THE QUATERNION WEIGHTED MOORE-PENROSE INVERSE AND ITS APPLICATIONS

Ivan I. Kyrchei *
Pidstrygach Institute for Applied Problems
of Mechanics and Mathematics of NAS of Ukraine,
Lviv, Ukraine

Abstract

The theory of noncommutative column-row determinants (previously introduced by the author) is extended to determinantal representations of the weighted Moore-Penrose inverse over the quaternion skew field in the chapter. To begin with, we introduce the weighted singular value decomposition (WSVD) of a quaternion matrix. Similarly as the singular value decomposition can be used for expressing the Moore-Penrose inverse, we give the representation of the weighted Moore-Penrose inverse by WSVD. Using this representation, limit and determinantal representations of the weighted Moore-Penrose inverse of a quaternion matrix are derived within the framework of the theory of column-row determinants. By using the obtained analogs of the adjoint matrix, we get the Cramer rules for the weighted Moore-Penrose solutions of left and right systems

*Corresponding Author Email: kyrchei@online.ua.

of quaternion linear equations, and for solutions of two-sided restricted quaternion matrix equation in all cases with respect to weighted matrices.
Numerical examples to illustrate the main results are given.

1. Introduction

Let $\mathbb{R}$ and $\mathbb{C}$ be the real and complex number fields, respectively. Throughout the paper, we denote the set of all $m \times n$ matrices over the quaternion skew field

$$\mathbb{H} = \{a_0 + a_1 i + a_2 j + a_3 k \mid i^2 = j^2 = k^2 = -1, \, a_0, a_1, a_2, a_3 \in \mathbb{R}\}$$

by $\mathbb{H}^{m\times n}$, and by $\mathbb{H}_r^{m\times n}$ the set of all $m \times n$ matrices over $\mathbb{H}$ with a rank r. Let $\mathrm{M}(n, \mathbb{H})$ be the ring of $n \times n$ quaternion matrices and $\mathbf{I}$ be the identity matrix with the appropriate size. For $\mathbf{A} \in \mathbb{H}^{n\times m}$, we denote by $\mathbf{A}^*$, $\operatorname{rank} \mathbf{A}$ the conjugate transpose (Hermitian adjoint) matrix and the rank of $\mathbf{A}$. The matrix $\mathbf{A} = (a_{ij}) \in \mathbb{H}^{n\times n}$ is Hermitian if $\mathbf{A}^* = \mathbf{A}$.

The definitions of the generalized inverse matrices can be extended to quaternion matrices as follows.

The Moore-Penrose inverse of $\mathbf{A} \in \mathbb{H}^{m\times n}$, denoted by $\mathbf{A}^\dagger$, is the unique matrix $\mathbf{X} \in \mathbb{H}^{n\times m}$ satisfying the following equations [1],

$$\mathbf{AXA} = \mathbf{A}; \tag{1}$$
$$\mathbf{XAX} = \mathbf{X}; \tag{2}$$
$$(\mathbf{AX})^* = \mathbf{AX}; \tag{3}$$
$$(\mathbf{XA})^* = \mathbf{XA}. \tag{4}$$

Let Hermitian positive definite matrices $\mathbf{M}$ and $\mathbf{N}$ of order m and n, respectively, be given. For $\mathbf{A} \in \mathbb{H}^{m\times n}$, **the weighted Moore-Penrose inverse** of $\mathbf{A}$ is the unique solution $\mathbf{X} = \mathbf{A}^\dagger_{M,N}$ of the matrix equations (1) and (2) and the following equations in $\mathbf{X}$ [2]:

$$(3M)\ (\mathbf{MAX})^* = \mathbf{MAX}; \ (4N)\ (\mathbf{NXA})^* = \mathbf{NXA}.$$

In particular, when $\mathbf{M} = \mathbf{I}_m$ and $\mathbf{N} = \mathbf{I}_n$, the matrix $\mathbf{X}$ satisfying the equations (1), (2), (3M), (4N) is the Moore-Penrose inverse $\mathbf{A}^\dagger$.

It is known various representations of the weighted Moore-Penrose. In particular, limit representations have been considered in [3, 4]. Determinantal representations of the complex (real) weighted Moore-Penrose have been derived by full-rank factorization in [5], by limit representation in [6] using the method first introduced in [7], and by minors in [8]. A basic method for finding the Moore-Penrose inverse is based on the singular value decomposition (SVD). It is available for quaternion matrices, (see, e.g. [9, 10]). The weighted Moore-Penrose inverse $\mathbf{A}^{\dagger}_{M,N} \in \mathbb{C}^{m \times n}$ can be explicitly expressed by the weighted singular value decomposition (WSVD) which at first has been obtained in [11] by Cholesky factorization. In [12], WSVD of real matrices with singular weights has been derived using weighted orthogonal matrices and weighted pseudoorthogonal matrices.

But why determinantal representations of generalized inverses are so important? When we return to the usual inverse, its determinantal representation is the matrix with cofactors in entries that gives direct method of its finding and makes it applicable in Cramer's rule for systems of linear equations. The same be wanted for generalized inverses. But there is not so unambiguous even for complex or real matrices. Therefore, there are various determinantal representations of generalized inverses because of looking of their explicit more applicable expressions (see, e.g. [13]).

The understanding of the problem for determinantal representing of generalized inverses as well as solutions and generalized inverse solutions of quaternion matrix equations, only now begins to be decided due to the theory of column-row determinants introduced in [10, 14]. Within the framework of the theory of noncommutative column-row determinants and using SVD of quaternion matrices, the limit and determinantal representations of the Moore-Penrose inverse over the quaternion skew field have been obtained in [15].

Song at al. [16, 17] have studied the weighted Moore-Penrose inverse over the quaternion skew field and obtained its determinantal representation within the framework of the theory of column-row determinants as well. But WSVD of quaternion matrices has not been considered and for obtaining a determinantal representation there was used auxiliary matrices which different from $\mathbf{A}$, and weights $\mathbf{M}$ and $\mathbf{N}$.

Weighted singular value decomposition (WSVD) and a representation of the weighted Moore-Penrose inverse of a quaternion matrix by WSVD recently have been derived by the author in [18]. The main goals of the chapter are introducing WSVD of quaternion matrices and representation of the weighted Moore-Penrose inverse over the quaternion skew field by WSVD, and then with their help, obtaining its limit and determinantal representations. As applications obtained determinantal representations, we give analogs of Cramer's rule for left and right systems of quaternion linear equations, the quaternion restricted matrix equation $\mathbf{AXB} = \mathbf{D}$, and consequently, $\mathbf{AX} = \mathbf{D}$ and $\mathbf{XB} = \mathbf{D}$.

It need to note that currently the theory of column-row determinants of quaternion matrices is active developing. Within the framework of column-row determinants, determinantal representations of various kind of generalized inverses, (generalized inverses) solutions of quaternion matrix equations recently have been derived as by the author (see, e.g. [19–23]) so by other researchers (see, e.g. [24–27]).

In this chapter we shall adopt the following notation.

Let $\alpha := \{\alpha_1, \ldots, \alpha_k\} \subseteq \{1, \ldots, m\}$ and $\beta := \{\beta_1, \ldots, \beta_k\} \subseteq \{1, \ldots, n\}$ be subsets of the order $1 \leq k \leq \min\{m, n\}$. By $\mathbf{A}_{\beta}^{\alpha}$ denote the submatrix of $\mathbf{A}$ determined by the rows indexed by α, and the columns indexed by β. Then, $\mathbf{A}_{\alpha}^{\alpha}$ denotes a principal submatrix determined by the rows and columns indexed by α. If $\mathbf{A} \in \mathrm{M}(n, \mathbb{H})$ is Hermitian, then by $|\mathbf{A}_{\alpha}^{\alpha}|$ denote the corresponding principal minor of $\det \mathbf{A}$, since $\mathbf{A}_{\alpha}^{\alpha}$ is Hermitian as well. For $1 \leq k \leq n$, denote by $L_{k,n} := \{\alpha : \alpha = (\alpha_1, \ldots, \alpha_k),\ 1 \leq \alpha_1 \leq \ldots \leq \alpha_k \leq n\}$ the collection of strictly increasing sequences of k integers chosen from $\{1, \ldots, n\}$. For fixed $i \in \alpha$ and $j \in \beta$, let

$$I_{r,m}\{i\} := \{\alpha : \alpha \in L_{r,m}, i \in \alpha\}, \quad J_{r,n}\{j\} := \{\beta : \beta \in L_{r,n}, j \in \beta\}.$$

The chapter is organized as follows. We start with some basic concepts and results from the theory of row-column determinants in Section 2. Some provisions of quaternion matrices and quaternion vector spaces are considered in Section 3. Weighted singular value decomposition and a representation of the weighted Moore-Penrose inverse of quaternion matrices by WSVD are considered in Subsection 4.1, and its limit representations in Subsection 4.2. In Section 5, we give the determinantal representations of the weighted Moore-Penrose inverse in both different cases. In Subsection 5.1, the matrices $\mathbf{N}^{-1}\mathbf{A}^*\mathbf{MA}$ and $\mathbf{AN}^{-1}\mathbf{A}^*\mathbf{M}$ are Hermitian, and they are non-Hermitian in Subsection 5.2. In Section 6, we obtain explicit representation formulas of the weighted Moore-Penrose solutions (analogs of Cramer's rule) of the left and right systems of linear equations over the quaternion skew field. In Section 7, we give analogs of Cramer's rule for the quaternion restricted matrix equation $\mathbf{AXB} = \mathbf{D}$, and consequently, $\mathbf{AX} = \mathbf{D}$ and $\mathbf{XB} = \mathbf{D}$. We consider all possible cases. In Section 8, we give numerical examples to illustrate the main results.

2. Preliminaries

For a quadratic matrix $\mathbf{A} = (a_{ij}) \in \mathrm{M}(n, \mathbb{H})$ can be define n row determinants and n column determinants as follows.

Suppose S_n is the symmetric group on the set $I_n = \{1, \ldots, n\}$.

Definition 2.1. *[14] The ith row determinant of $\mathbf{A} = (a_{ij}) \in \mathrm{M}(n, \mathbb{H})$ is defined for all $i = 1, \ldots, n$ by putting*

$$\mathrm{rdet}_i \mathbf{A} = \sum_{\sigma \in S_n} (-1)^{n-r} (a_{i\, i_{k_1}} a_{i_{k_1} i_{k_1+1}} \ldots a_{i_{k_1+l_1} i}) \ldots (a_{i_{k_r} i_{k_r+1}} \ldots a_{i_{k_r+l_r} i_{k_r}}),$$

$$\sigma = (i\, i_{k_1} i_{k_1+1} \ldots i_{k_1+l_1})(i_{k_2} i_{k_2+1} \ldots i_{k_2+l_2}) \ldots (i_{k_r} i_{k_r+1} \ldots i_{k_r+l_r}),$$

with conditions $i_{k_2} < i_{k_3} < \ldots < i_{k_r}$ and $i_{k_t} < i_{k_t+s}$ for $t = 2, \ldots, r$ and $s = 1, \ldots, l_t$.

Definition 2.2. *[14] The jth column determinant of $\mathbf{A} = (a_{ij}) \in \mathrm{M}(n, \mathbb{H})$ is defined for all $j = 1, \ldots, n$ by putting*

$$\mathrm{cdet}_j \mathbf{A} = \sum_{\tau \in S_n} (-1)^{n-r} (a_{j_{k_r} j_{k_r+l_r}} \ldots a_{j_{k_r+1} i_{k_r}}) \ldots (a_{j\, j_{k_1+l_1}} \ldots a_{j_{k_1+1} j_{k_1}} a_{j_{k_1} j}),$$

$$\tau = (j_{k_r+l_r} \ldots j_{k_r+1} j_{k_r}) \ldots (j_{k_2+l_2} \ldots j_{k_2+1} j_{k_2}) (j_{k_1+l_1} \ldots j_{k_1+1} j_{k_1} j),$$

with conditions, $j_{k_2} < j_{k_3} < \ldots < j_{k_r}$ and $j_{k_t} < j_{k_t+s}$ for $t = 2, \ldots, r$ and $s = 1, \ldots, l_t$.

Suppose $\mathbf{A}^{ij}$ denotes the submatrix of $\mathbf{A}$ obtained by deleting both the ith row and the jth column. Let $\mathbf{a}_{.j}$ be the jth column and $\mathbf{a}_{i.}$ be the ith row of $\mathbf{A}$. Suppose $\mathbf{A}_{.j}(\mathbf{b})$ denotes the matrix obtained from $\mathbf{A}$ by replacing its jth column with the column $\mathbf{b}$, and $\mathbf{A}_{i.}(\mathbf{b})$ denotes the matrix obtained from $\mathbf{A}$ by replacing its ith row with the row $\mathbf{b}$. We note some properties of column and row determinants of a quaternion matrix $\mathbf{A} = (a_{ij})$, where $i \in I_n$, $j \in J_n$ and $I_n = J_n = \{1, \ldots, n\}$.

Proposition 2.1. *[14] If $b \in \mathbb{H}$, then*

$$\mathrm{rdet}_i \mathbf{A}_{i.}(b \cdot \mathbf{a}_{i.}) = b \cdot \mathrm{rdet}_i \mathbf{A},$$
$$\mathrm{cdet}_i \mathbf{A}_{.i}(\mathbf{a}_{.i} \cdot b) = \mathrm{cdet}_i \mathbf{A} \cdot b,$$

for all $i = 1, \ldots, n$.

Proposition 2.2. *[14] If for* $\mathbf{A} \in \mathrm{M}(n, \mathbb{H})$ *there exists* $t \in I_n$ *such that* $a_{tj} = b_j + c_j$ *for all* $j = 1, \ldots, n$*, then*

$$\mathrm{rdet}_i\, \mathbf{A} = \mathrm{rdet}_i\, \mathbf{A}_{t.}(\mathbf{b}) + \mathrm{rdet}_i\, \mathbf{A}_{t.}(\mathbf{c}),$$
$$\mathrm{cdet}_i\, \mathbf{A} = \mathrm{cdet}_i\, \mathbf{A}_{t.}(\mathbf{b}) + \mathrm{cdet}_i\, \mathbf{A}_{t.}(\mathbf{c}),$$

where $\mathbf{b} = (b_1, \ldots, b_n)$, $\mathbf{c} = (c_1, \ldots, c_n)$ *and for all* $i = 1, \ldots, n$.

Proposition 2.3. *[14] If for* $\mathbf{A} \in \mathrm{M}(n, \mathbb{H})$ *there exists* $t \in J_n$ *such that* $a_{i\,t} = b_i + c_i$ *for all* $i = 1, \ldots, n$*, then*

$$\mathrm{rdet}_j\, \mathbf{A} = \mathrm{rdet}_j\, \mathbf{A}_{.\,t}(\mathbf{b}) + \mathrm{rdet}_j\, \mathbf{A}_{.\,t}(\mathbf{c}),$$
$$\mathrm{cdet}_j\, \mathbf{A} = \mathrm{cdet}_j\, \mathbf{A}_{.\,t}(\mathbf{b}) + \mathrm{cdet}_j \mathbf{A}_{.\,t}(\mathbf{c}),$$

where $\mathbf{b} = (b_1, \ldots, b_n)^T$, $\mathbf{c} = (c_1, \ldots, c_n)^T$ *and for all* $j = 1, \ldots, n$.

Remark 2.1. *Let* $\mathrm{rdet}_i\, \mathbf{A} = \sum\limits_{j=1}^{n} a_{ij} \cdot R_{ij}$ *and* $\mathrm{cdet}_j\, \mathbf{A} = \sum\limits_{i=1}^{n} L_{ij} \cdot a_{ij}$ *for all* $i, j = 1, \ldots, n$*, where by* R_{ij} *and* L_{ij} *denote the right and left* (ij)*th cofactors of* $\mathbf{A} \in \mathrm{M}(n, \mathbb{H})$*, respectively. It means that* $\mathrm{rdet}_i\, \mathbf{A}$ *can be expand by right cofactors along the* i*th row and* $\mathrm{cdet}_j \mathbf{A}$ *can be expand by left cofactors along the* j*th column, respectively, for all* $i, j = 1, \ldots, n$.

The main property of the usual determinant is that the determinant of a non-invertible matrix must be equal zero. But the row and column determinants don't satisfy it, in general. Therefore, these matrix functions can be consider as some pre-determinants. The following theorem has a key value in the theory of the column and row determinants.

Theorem 2.1. *[14] If* $\mathbf{A} = (a_{ij}) \in \mathrm{M}(n, \mathbb{H})$ *is Hermitian, then* $\mathrm{rdet}_1 \mathbf{A} = \cdots = \mathrm{rdet}_n \mathbf{A} = \mathrm{cdet}_1 \mathbf{A} = \cdots = \mathrm{cdet}_n \mathbf{A} \in \mathbb{R}$.

Due to Theorem 2.1, we can define the determinant of a Hermitian matrix $\mathbf{A} \in \mathrm{M}(n, \mathbb{H})$ by putting,

$$\det \mathbf{A} := \mathrm{rdet}_i\, \mathbf{A} = \mathrm{cdet}_i\, \mathbf{A}, \tag{5}$$

for all $i = 1, \ldots, n$. By using its row and column determinants, the determinant of a quaternion Hermitian matrix has properties similar to the usual determinant. These properties are completely explored in [14] and can be summarized in the following theorems.

Theorem 2.2. *If the ith row of a Hermitian matrix $\mathbf{A} \in \mathrm{M}(n, \mathbb{H})$ is replaced with a left linear combination of its other rows, i.e. $\mathbf{a}_{i.} = c_1\mathbf{a}_{i_1.} + \ldots + c_k\mathbf{a}_{i_k.}$, where $c_l \in \mathbb{H}$ for all $l = 1, \ldots, k$ and $\{i, i_l\} \subset I_n$, then*

$$\mathrm{rdet}_i\, \mathbf{A}_{i.}\left(c_1\mathbf{a}_{i_1.} + \ldots + c_k\mathbf{a}_{i_k.}\right) = \mathrm{cdet}_i\, \mathbf{A}_{i.}\left(c_1\mathbf{a}_{i_1.} + \ldots + c_k\mathbf{a}_{i_k.}\right) = 0.$$

Theorem 2.3. *If the jth column of a Hermitian matrix $\mathbf{A} \in \mathrm{M}(n, \mathbb{H})$ is replaced with a right linear combination of its other columns, i.e. $\mathbf{a}_{.j} = \mathbf{a}_{.j_1}c_1 + \ldots + \mathbf{a}_{.j_k}c_k$, where $c_l \in \mathbb{H}$ for all $l = 1, \ldots, k$ and $\{j, j_l\} \subset J_n$, then*

$$\mathrm{cdet}_j\, \mathbf{A}_{.j}\left(\mathbf{a}_{.j_1}c_1 + \ldots + \mathbf{a}_{.j_k}c_k\right) = \mathrm{rdet}_j\, \mathbf{A}_{.j}\left(\mathbf{a}_{.j_1}c_1 + \ldots + \mathbf{a}_{.j_k}c_k\right) = 0.$$

The following theorem about determinantal representation of an inverse matrix of Hermitian follows immediately from these properties.

Theorem 2.4. *[14] If a Hermitian matrix $\mathbf{A} \in \mathrm{M}(n, \mathbb{H})$ is such that $\det \mathbf{A} \neq 0$, then there exist a unique right inverse matrix $(R\mathbf{A})^{-1}$ and a unique left inverse matrix $(L\mathbf{A})^{-1}$, and $(R\mathbf{A})^{-1} = (L\mathbf{A})^{-1} =: \mathbf{A}^{-1}$, which possess the following determinantal representations:*

$$(R\mathbf{A})^{-1} = \frac{1}{\det \mathbf{A}} \begin{pmatrix} R_{11} & R_{21} & \cdots & R_{n1} \\ R_{12} & R_{22} & \cdots & R_{n2} \\ \cdots & \cdots & \cdots & \cdots \\ R_{1n} & R_{2n} & \cdots & R_{nn} \end{pmatrix}, \tag{6}$$

$$(L\mathbf{A})^{-1} = \frac{1}{\det \mathbf{A}} \begin{pmatrix} L_{11} & L_{21} & \cdots & L_{n1} \\ L_{12} & L_{22} & \cdots & L_{n2} \\ \cdots & \cdots & \cdots & \cdots \\ L_{1n} & L_{2n} & \cdots & L_{nn} \end{pmatrix}, \tag{7}$$

where $\det \mathbf{A} = \sum\limits_{j=1}^{n} a_{ij} \cdot R_{ij} = \sum\limits_{i=1}^{n} L_{ij} \cdot a_{ij}$,

$$R_{ij} = \begin{cases} -\mathrm{rdet}_j\, \mathbf{A}_{.j}^{ii}\left(\mathbf{a}_{.i}\right), & i \neq j, \\ \mathrm{rdet}_k\, \mathbf{A}^{ii}, & i = j, \end{cases} \quad L_{ij} = \begin{cases} -\mathrm{cdet}_i\, \mathbf{A}_{i.}^{jj}\left(\mathbf{a}_{j.}\right), & i \neq j, \\ \mathrm{cdet}_k\, \mathbf{A}^{jj}, & i = j. \end{cases}$$

The submatrix $\mathbf{A}_{.j}^{ii}\left(\mathbf{a}_{.i}\right)$ is obtained from $\mathbf{A}$ by replacing the jth column with the ith column and then deleting both the ith row and column, $\mathbf{A}_{i.}^{jj}\left(\mathbf{a}_{j.}\right)$ is obtained by replacing the ith row with the jth row, and then by deleting both the jth row and column, respectively. $I_n = \{1, \ldots, n\}$, $k = \min\{I_n \setminus \{i\}\}$, for all $i, j = 1, \ldots, n$.

Theorem 2.5. *[14] If an arbitrary column of* $\mathbf{A} \in \mathbf{H}^{m\times n}$ *is a right linear combination of its other columns, or an arbitrary row of* $\mathbf{A}^*$ *is a left linear combination of its others, then* $\det \mathbf{A}^*\mathbf{A} = 0$.

Theorem 2.6. *[14] The right-linearly independence of columns of* $\mathbf{A} \in \mathbf{H}^{m\times n}$ *or the left-linearly independence of rows of* $\mathbf{A}^*$ *is the necessary and sufficient condition for* $\det \mathbf{A}^*\mathbf{A} \neq 0$.

Theorem 2.7. *If* $\mathbf{A} \in \mathrm{M}(n, \mathbb{H})$, *then* $\det \mathbf{A}\mathbf{A}^* = \det \mathbf{A}^*\mathbf{A}$.

Definition 2.3. *For* $\mathbf{A} \in \mathrm{M}(n, \mathbb{H})$, *the double determinant of* $\mathbf{A}$ *is defined by putting,* $\mathrm{ddet}\, \mathbf{A} := \det \mathbf{A}\mathbf{A}^* = \det \mathbf{A}^*\mathbf{A}$.

Theorem 2.8. *The necessary and sufficient condition of invertibility of* $\mathbf{A} \in \mathrm{M}(n, \mathbb{H})$ *is* $\mathrm{ddet}\mathbf{A} \neq 0$. *Then there exists* $\mathbf{A}^{-1} = (L\mathbf{A})^{-1} = (R\mathbf{A})^{-1}$, *where*

$$(L\mathbf{A})^{-1} = (\mathbf{A}^*\mathbf{A})^{-1}\,\mathbf{A}^* = \frac{1}{\mathrm{ddet}\mathbf{A}}\begin{pmatrix} \mathbb{L}_{11} & \mathbb{L}_{21} & \dots & \mathbb{L}_{n1} \\ \mathbb{L}_{12} & \mathbb{L}_{22} & \dots & \mathbb{L}_{n2} \\ \dots & \dots & \dots & \dots \\ \mathbb{L}_{1n} & \mathbb{L}_{2n} & \dots & \mathbb{L}_{nn} \end{pmatrix}$$

$$(R\mathbf{A})^{-1} = \mathbf{A}^*\,(\mathbf{A}\mathbf{A}^*)^{-1} = \frac{1}{\mathrm{ddet}\mathbf{A}^*}\begin{pmatrix} \mathbb{R}_{11} & \mathbb{R}_{21} & \dots & \mathbb{R}_{n1} \\ \mathbb{R}_{12} & \mathbb{R}_{22} & \dots & \mathbb{R}_{n2} \\ \dots & \dots & \dots & \dots \\ \mathbb{R}_{1n} & \mathbb{R}_{2n} & \dots & \mathbb{R}_{nn} \end{pmatrix}$$

and

$$\mathbb{L}_{ij} = \mathrm{cdet}_j(\mathbf{A}^*\mathbf{A})_{.j}\left(\mathbf{a}^*_{.i}\right),\quad \mathbb{R}_{ij} = \mathrm{rdet}_i(\mathbf{A}\mathbf{A}^*)_{i.}\left(\mathbf{a}^*_{j.}\right),$$

for all $i, j = 1, \dots, n$.

Moreover, the following criterion of invertibility of a quaternion matrix can be obtained.

Theorem 2.9. *If* $\mathbf{A} \in \mathrm{M}(n, \mathbb{H})$, *then the following statements are equivalent.*

i) $\mathbf{A}$ *is invertible, i.e.* $\mathbf{A} \in GL(n, \mathbb{H})$;

ii) rows of $\mathbf{A}$ *are left-linearly independent;*

iii) columns of $\mathbf{A}$ *are right-linearly independent;*

iv) $\mathrm{ddet}\, \mathbf{A} \neq 0$.

2.1. Some Provisions of Quaternion Matrices and Quaternion Vector Spaces

Due to real-scalar multiplying on the right, quaternion column-vectors form a right vector $\mathbb{R}$-space, and, by real-scalar multiplying on the left, quaternion row-vectors form a left vector $\mathbb{R}$-space. Moreover, we define right and left quaternion vector spaces, denoted by $\mathcal{H}_r$ and $\mathcal{H}_l$, respectively, with corresponding $\mathbb{H}$-valued inner products $\langle\cdot,\cdot\rangle$ which satisfy, for every $\alpha,\beta\in\mathbb{H}$, and $\mathbf{x},\mathbf{y},\mathbf{z}\in\mathcal{H}_r(\mathcal{H}_l)$, the relations:

$$\langle\mathbf{x},\mathbf{y}\rangle=\overline{\langle\mathbf{y},\mathbf{x}\rangle};$$

$$\langle\mathbf{x},\mathbf{x}\rangle\geq 0\in\mathbb{R}\ \text{ and }\ \|\mathbf{x}\|^2:=\langle\mathbf{x},\mathbf{x}\rangle=0\ \Leftrightarrow\ \mathbf{x}=\mathbf{0};$$

$$\begin{array}{ll}\langle\mathbf{x}\alpha+\mathbf{y}\beta,\mathbf{z}\rangle=\langle\mathbf{x},\mathbf{z}\rangle\alpha+\langle\mathbf{y},\mathbf{z}\rangle\beta & \text{when } \mathbf{x},\mathbf{y},\mathbf{z}\in\mathcal{H}_r\\ \langle\alpha\mathbf{x}+\beta\mathbf{y},\mathbf{z}\rangle=\alpha\langle\mathbf{x},\mathbf{z}\rangle+\beta\langle\mathbf{y},\mathbf{z}\rangle & \text{when } \mathbf{x},\mathbf{y},\mathbf{z}\in\mathcal{H}_l;\end{array}$$

$$\begin{array}{ll}\langle\mathbf{x},\mathbf{y}\alpha+\mathbf{z}\beta\rangle=\overline{\alpha}\langle\mathbf{x},\mathbf{y}\rangle+\overline{\beta}\langle\mathbf{x},\mathbf{z}\rangle & \text{when } \mathbf{x},\mathbf{y},\mathbf{z}\in\mathcal{H}_r\\ \langle\mathbf{x},\alpha\mathbf{y}+\beta\mathbf{z}\rangle=\langle\mathbf{x},\mathbf{y}\rangle\overline{\alpha}+\langle\mathbf{y},\mathbf{z}\rangle\overline{\beta} & \text{when } \mathbf{x},\mathbf{y},\mathbf{z}\in\mathcal{H}_l.\end{array}$$

It can be achieved by putting $\langle\mathbf{x},\mathbf{y}\rangle_r=\overline{y}_1x_1+\cdots+\overline{y}_nx_n$ for $\mathbf{x}=(x_i)_{i=1}^n,\mathbf{y}=(y_i)_{i=1}^n\in\mathcal{H}_r$, and $\langle\mathbf{x},\mathbf{y}\rangle_l=x_1\overline{y}_1+\cdots+x_n\overline{y}_n$ for $\mathbf{x},\mathbf{y}\in\mathcal{H}_l$.

The right vector spaces $\mathcal{H}_r$ possess the Gram-Schmidt process which takes a nonorthogonal set of linearly independent vectors $S=\{\mathbf{v}_1,...,\mathbf{v}_k\}$ for $k\leq n$ and constructs an orthogonal (or orthonormal) basis $S^{'}=\{\mathbf{u}_1,...,\mathbf{u}_k\}$ that spans the same k-dimensional subspace of $\mathcal{H}_r$ as S. To $\mathcal{H}_r$, the following projection operator is defined by

$$\mathrm{proj}_{\mathbf{u}}(\mathbf{v}):=\mathbf{u}\frac{\langle\mathbf{u},\mathbf{v}\rangle_r}{\langle\mathbf{u},\mathbf{u}\rangle_r},$$

which orthogonally projects the vector $\mathbf{v}$ onto the line spanned by the vector $\mathbf{u}$. Then, the GramSchmidt process works as follows:

$$\begin{array}{ll}\mathbf{u}_1=\mathbf{v}_1, & \mathbf{e}_1=\frac{\mathbf{u}_1}{\|\mathbf{u}_1\|},\\ \mathbf{u}_k=\mathbf{v}_k-\sum_{j=1}^{k-1}\mathrm{proj}_{\mathbf{u_j}}(\mathbf{v_k}), & \mathbf{e}_k=\frac{\mathbf{u}_k}{\|\mathbf{u}_k\|}.\end{array}$$

The sequence $\mathbf{u}_1,...,\mathbf{u}_k$ is the required system of orthogonal vectors, and the normalized vectors $\mathbf{e}_1,...,\mathbf{e}_k$ form an orthonormal set.

The GramSchmidt process for the left vector spaces $\mathcal{H}_l$ can be realize by the same algorithm but with the projection operator

$$\mathrm{proj}_{\mathbf{u}}(\mathbf{v}) := \frac{\langle \mathbf{u}, \mathbf{v} \rangle_l}{\langle \mathbf{u}, \mathbf{u} \rangle_l} \mathbf{u}.$$

Definition 2.4. *Suppose* $\mathbf{U} \in \mathrm{M}\,(n, \mathbb{H})$ *and* $\mathbf{U}^*\mathbf{U} = \mathbf{U}\mathbf{U}^* = \mathbf{I}$, *then the matrix* $\mathbf{U}$ *is called unitary.*

Clear, that columns of $\mathbf{U}$ form a system of normalized vectors in $\mathcal{H}_r$, rows of $\mathbf{U}^*$ is a system of normalized vectors in $\mathcal{H}_l$.

We shall also need the following facts about eigenvalues of quaternion matrices.

Due to the noncommutativity of quaternions, there are two types of eigenvalues. A quaternion λ is said to be a left eigenvalue of $\mathbf{A} \in \mathrm{M}\,(n, \mathbb{H})$ if

$$\mathbf{A} \cdot \mathbf{x} = \lambda \cdot \mathbf{x}, \tag{8}$$

and a right eigenvalue if

$$\mathbf{A} \cdot \mathbf{x} = \mathbf{x} \cdot \lambda \tag{9}$$

for some nonzero quaternion column-vector $\mathbf{x}$. Then, the set $\{\lambda \in \mathbb{H} | \mathbf{A}\mathbf{x} = \lambda\mathbf{x},\ \mathbf{x} \neq \mathbf{0} \in \mathbb{H}^n\}$ is called the left spectrum of $\mathbf{A}$, denoted by $\sigma_l(\mathbf{A})$. The right spectrum is similarly defined by putting, $\sigma_r(\mathbf{A}) := \{\lambda \in \mathbb{H} | \mathbf{A}\mathbf{x} = \mathbf{x}\lambda,\ \mathbf{x} \neq \mathbf{0} \in \mathbb{H}^n\}$.

The theory on the left eigenvalues of quaternion matrices has been investigated in particular in [29–31]. The theory on the right eigenvalues of quaternion matrices is more developed [32–37]. We consider this is a natural consequence of the fact that quaternion column vectors form a right vector space for which left eigenvalues from (8) seem to be "exotic" because of their multiplying from the left. Left eigenvalues may appear natural in the equation

$$\mathbf{x}\mathbf{A} = \lambda\mathbf{x}. \tag{10}$$

Since $\mathbf{x}\mathbf{A} = \lambda\mathbf{x}$ if and only if $\mathbf{A}^*\mathbf{x}^* = \mathbf{x}^*\overline{\lambda}$, then the theory of such "natural" left eigenvalues from (10) be identical to the theory of right eigenvalues from (9). Similarly, the theory of right eigenvalues from $\mathbf{x}\mathbf{A} = \mathbf{x}\lambda$ is identical to the theory of left eigenvalues from (8).

Now, we present the some known results from the theory of right eigenvalues (9) that will be applied hereinafter. Due to the above, henceforth, we will avoid the "right" specification.

In particular, it's well known that if λ is a nonreal eigenvalue of $\mathbf{A}$, so is any element in the equivalence class containing $[\lambda]$, i.e. $[\lambda] = \{x | x = u^{-1}\lambda u,\ u \in \mathbb{H},\ \|u\| = 1\}$.

Theorem 2.10. *[32] Any quaternion matrix $\mathbf{A} \in \mathrm{M}(n, \mathbb{H})$ has exactly n eigenvalues which are complex numbers with nonnegative imaginary parts.*

Proposition 2.4. *[34] Suppose that $\lambda_1, \dots, \lambda_n$ are distinct eigenvalues for $\mathbf{A} \in \mathrm{M}(n, \mathbb{H})$, no two of which are conjugate, and let $\mathbf{v}_1, \dots, \mathbf{v}_n$ be corresponding eigenvectors. Then $\mathbf{v}_1, \dots, \mathbf{v}_n$ are (right) linearly independent.*

Moreover, similarly to the complex case, the following theorem can be proved.

Theorem 2.11. *A matrix $\mathbf{A} \in \mathrm{M}(n, \mathbb{H})$ is diagonalizable if and only if $\mathbf{A}$ has a set of n right-linearly independent eigenvectors. Furthermore, if λ_i, $\mathbf{v}_1$, for $i = 1, \dots, n$, are eigenpairs of $\mathbf{A}$, then*

$$\mathbf{A} = \mathbf{P}\mathbf{D}\mathbf{P}^{-1},$$

where $\mathbf{P} = [\mathbf{v}_1, \dots, \mathbf{v}_n]$, $\mathbf{D} = diag\,[\lambda_1, \dots, \lambda_n]$.

Proof. ($\Rightarrow$) Since matrix $\mathbf{A}$ is diagonalizable, there exist an invertible matrix $\mathbf{P}$ and a diagonal matrix $\mathbf{D}$ such that $\mathbf{A} = \mathbf{P}\mathbf{D}\mathbf{P}^{-1}$. Then,

$$\mathbf{D} = \mathbf{P}^{-1}\mathbf{A}\mathbf{P}. \tag{11}$$

Since $\mathbf{D} \in \mathrm{M}(n, \mathbb{H})$ is diagonal, it has a right-linearly independent set of n right eigenvectors, given by the column vectors of the identity matrix, i.e.,

$$\mathbf{D}\mathbf{e}_i = \mathbf{e}_i d_{ii},\ \ \mathbf{D} = diag\,[d_{11}, \dots, d_{nn}]\,,\ \ \mathbf{I} = [\mathbf{e}_1, \dots, \mathbf{e}_n]\,.$$

So, the pair d_{ii}, $\mathbf{e}_i$ is an eigenvalue-eigenvector pair of $\mathbf{D}$, for all $i = 1, \dots, n$. Due to $\mathbf{e}_i d_{ii} = d_{ii}\mathbf{e}_i$ and using (11), we have

$$\mathbf{e}_i d_{ii} = d_{ii}\mathbf{e}_i = \mathbf{D}\mathbf{e}_i = \mathbf{P}^{-1}\mathbf{A}\mathbf{P}\mathbf{e}_i. \tag{12}$$

By multiplying the extreme members of (12) by $\mathbf{P}$ on the left, we obtain

$$\mathbf{A}(\mathbf{P}\mathbf{e}_i) = (\mathbf{P}\mathbf{e}_i)d_{ii}.$$

It means that the vectors $\mathbf{v}_i = \mathbf{P}\mathbf{e}_i$ are right eigenvectors of $\mathbf{A}$ with eigenvalue d_{ii} for all $i = 1, \ldots, n$. Since the matrix $\mathbf{P} = [\mathbf{v}_1, \ldots, \mathbf{v}_n]$ is invertible, then, by Theorem 2.9, the eigenvectors $\mathbf{v}_1, \ldots, \mathbf{v}_n$ is right-linearly independent.

($\Leftarrow$) Let λ_i, $\mathbf{v}_1$, be eigenvalue-eigenvector pairs of $\mathbf{A}$, for $i = 1, \ldots, n$. Consider the matrix $\mathbf{P} = [\mathbf{v}_1, \ldots, \mathbf{v}_n]$. Computing the product, we obtain

$$\mathbf{AP} = \mathbf{A}\left[\mathbf{v}_1, \ldots, \mathbf{v}_n\right] = \left[\mathbf{A}\mathbf{v}_1, \ldots, \mathbf{A}\mathbf{v}_n\right] = \left[\mathbf{v}_1\lambda_1, \ldots, \mathbf{v}_n\lambda_n\right].$$

Since the eigenvector set $\{\mathbf{v}_1, \ldots, \mathbf{v}_n\}$ is right-linearly independent, then, by Theorem 2.9, $\mathbf{P}$ is invertible. There exists $\mathbf{P}^{-1}$, and

$$\mathbf{P}^{-1}\mathbf{AP} = \mathbf{P}^{-1}\left[\mathbf{v}_1\lambda_1, \ldots, \mathbf{v}_n\lambda_n\right] = \left[\mathbf{P}^{-1}\mathbf{v}_1\lambda_1, \ldots, \mathbf{P}^{-1}\mathbf{v}_n\lambda_n\right].$$

Since $\mathbf{P}^{-1}\mathbf{P} = \mathbf{I}$, then $\mathbf{P}^{-1}\mathbf{v}_i = \mathbf{e}_i$ for all $i = 1, \ldots, n$. So,

$$\mathbf{P}^{-1}\mathbf{AP} = \left[\mathbf{e}_1\lambda_1, \ldots, \mathbf{e}_n\lambda_n\right] = diag\left[\lambda_1, \ldots, \lambda_n\right]$$

Denoting $\mathbf{D} = diag\left[\lambda_1, \ldots, \lambda_n\right]$, we conclude that $\mathbf{P}^{-1}\mathbf{AP} = \mathbf{D}$, or equivalently, $\mathbf{A} = \mathbf{PDP}^{-1}$. □

Corollary 2.1. *[34]If $\mathbf{A} \in \mathrm{M}(n, \mathbb{H})$ has n non-conjugate eigenvalues, then it can be diagonalized in the sense that there is a $\mathbf{P} \in GL_n(\mathbb{H})$ for which $\mathbf{PAP}^{-1}$ is diagonal.*

Those eigenvalues $h_1 + k_1\mathbf{i}, \ldots, h_n + k_n\mathbf{i}$, where $k_t \geq 0$ and $h_t, k_t \in \mathbb{R}$ for all $t = 1, \ldots, n$, are said to be the standard eigenvalues of $\mathbf{A}$.

Theorem 2.12. *[32] Let $\mathbf{A} \in \mathrm{M}(n, \mathbb{H})$. Then there exists a unitary matrix $\mathbf{U}$ such that $\mathbf{U}^*\mathbf{AU}$ is an upper triangular matrix with diagonal entries $h_1 + k_1\mathbf{i}, \ldots, h_n + k_n\mathbf{i}$ which are the standard eigenvalues of $\mathbf{A}$.*

Corollary 2.2. *[36] Let $\mathbf{A} \in \mathrm{M}(n, \mathbb{H})$ with the standard eigenvalues $h_1 + k_1\mathbf{i}, \ldots, h_n + k_n\mathbf{i}$. Then $\sigma_r = [h_1 + k_1\mathbf{i}] \cup \cdots \cup [h_n + k_n\mathbf{i}]$.*

Corollary 2.3. *[36] $\mathbf{A} \in \mathrm{M}(n, \mathbb{H})$ is normal if and only if there exists an unitary matrix $\mathbf{U} \in \mathrm{M}(n, \mathbb{H})$ such that*

$$\mathbf{U}^*\mathbf{AU} = \mathrm{diag}\{\lambda_1, \ldots, \lambda_n\},$$

where $\lambda_i = h_i + k_i\mathbf{i} \in \mathbb{C}$ is standard eigenvalues for all $i = 1, \ldots, n$.

Corollary 2.4. *Let* $\mathbf{A} \in \mathrm{M}(n, \mathbb{H})$ *be given. Then,* $\mathbf{A}$ *is Hermitian if and only if there are a unitary matrix* $\mathbf{U} \in \mathrm{M}(n, \mathbb{H})$ *and a real diagonal matrix* $\mathbf{D} = \mathrm{diag}(\lambda_1, \lambda_2, \ldots, \lambda_n)$ *such that* $\mathbf{A} = \mathbf{U}\mathbf{D}\mathbf{U}^*$*, where* $\lambda_1, ..., \lambda_n$ *are right eigenvalues of* $\mathbf{A}$.

The right (9) and left (8) eigenvalues are in general unrelated [38], but it is not for Hermitian matrices. Suppose $\mathbf{A} \in \mathrm{M}(n, \mathbb{H})$ is Hermitian and $\lambda \in \mathbb{R}$ is its right eigenvalue, then $\mathbf{A} \cdot \mathbf{x} = \mathbf{x} \cdot \lambda = \lambda \cdot \mathbf{x}$. This means that all right eigenvalues of a Hermitian matrix are its left eigenvalues as well. For real left eigenvalues, $\lambda \in \mathbb{R}$, the matrix $\lambda\mathbf{I} - \mathbf{A}$ is Hermitian.

Definition 2.5. *If* $t \in \mathbb{R}$*, then for a Hermitian matrix* $\mathbf{A}$ *the polynomial* $p_{\mathbf{A}}(t) = \det(t\mathbf{I} - \mathbf{A})$ *is said to be the characteristic polynomial of* $\mathbf{A}$.

Lemma 2.1. *[14] If* $\mathbf{A} \in \mathrm{M}(n, \mathbb{H})$ *is Hermitian, then* $p_{\mathbf{A}}(t) = t^n - d_1 t^{n-1} + d_2 t^{n-2} - \ldots + (-1)^n d_n$*, where* d_k *is the sum of principle minors of* $\mathbf{A}$ *of order* k*,* $1 \leq k < n$*, and* $d_n = \det \mathbf{A}$.

The roots of the characteristic polynomial of a Hermitian matrix are its real left eigenvalues, which are its right eigenvalues as well.

Definition 2.6. *Let* $\mathbf{A} \in \mathbb{H}^{n \times n}$ *be a Hermitian matrix and* $\pi(\mathbf{A}) = \pi$ *be the number of positive eigenvalues of* $\mathbf{A}$*,* $\nu(\mathbf{A}) = \nu$ *be the number of negative eigenvalues of* $\mathbf{A}$*, and* $\delta(\mathbf{A}) = \delta$ *be the number of zero eigenvalues of* $\mathbf{A}$*. Then the ordered triple* $\omega = (\pi, \nu, \delta)$ *will be called the inertia of* $\mathbf{A}$*. We shall write* $\omega = In\,\mathbf{A}$.

We have [9, 10] the following theorem about the singular value decomposition (SVD) of quaternion matrices and their Moore-Penrose inverses.

Theorem 2.13. *(SVD) Let* $\mathbf{A} \in \mathbb{H}_r^{m \times n}$*. There exist unitary quaternion matrices* $\mathbf{V} \in \mathbb{H}^{m \times m}$ *and* $\mathbf{W} \in \mathbb{H}^{n \times n}$ *such that* $\mathbf{A} = \mathbf{V}\mathbf{\Sigma}\mathbf{W}^*$*, where* $\mathbf{\Sigma} = \begin{pmatrix} \mathbf{D} & \mathbf{0} \\ \mathbf{0} & \mathbf{0} \end{pmatrix} \in \mathbb{H}_r^{m \times n}$*, and* $\mathbf{D} = \mathrm{diag}(\sigma_1, \sigma_2, \ldots, \sigma_r), \sigma_1 \geq \sigma_2 \geq \ldots \geq \sigma_r > 0$*, and* σ_i^2 *is the nonzero eigenvalues of* $\mathbf{A}^*\mathbf{A}$ *for all* $i = 1, ..., r$*. Then* $\mathbf{A}^\dagger = \mathbf{W}\mathbf{\Sigma}^\dagger\mathbf{V}^*$*, where* $\mathbf{\Sigma}^\dagger = \begin{pmatrix} \mathbf{D}^{-1} & \mathbf{0} \\ \mathbf{0} & \mathbf{0} \end{pmatrix} \in \mathbb{H}_r^{n \times m}$.

Due to Theorem 2.13, the limit and determinantal representations of the Moore-Penrose inverse over the quaternion skew field have been obtained as follows.

Lemma 2.2. *[14] If* $\mathbf{A} \in \mathbb{H}^{m\times n}$ *and* $\mathbf{A}^\dagger$ *is its Moore-Penrose inverse, then* $\mathbf{A}^\dagger = \lim\limits_{\alpha\to 0} \mathbf{A}^* \left(\mathbf{A}\mathbf{A}^* + \alpha \mathbf{I}\right)^{-1} = \lim\limits_{\alpha\to 0} \left(\mathbf{A}^*\mathbf{A} + \alpha \mathbf{I}\right)^{-1} \mathbf{A}^*$, *where* $\alpha \in \mathbb{R}_\dagger$.

Theorem 2.14. *[14] If* $\mathbf{A} \in \mathbb{H}_r^{m\times n}$, *then the Moore-Penrose inverse* $\mathbf{A}^\dagger = \left(a_{ij}^\dagger\right) \in \mathbb{H}^{n\times m}$ *possess the following determinantal representations:*

$$a_{ij}^\dagger = \frac{\sum\limits_{\beta\in J_{r,\,n}\{i\}} \operatorname{cdet}_i \left(\left(\mathbf{A}^*\mathbf{A}\right)_{.i} \left(\mathbf{a}_{.j}^*\right)\right) {}_\beta^\beta}{\sum\limits_{\beta\in J_{r,\,n}} \left|\left(\mathbf{A}^*\mathbf{A}\right) {}_\beta^\beta\right|}, \tag{13}$$

or

$$a_{ij}^\dagger = \frac{\sum\limits_{\alpha\in I_{r,m}\{j\}} \operatorname{rdet}_j \left(\left(\mathbf{A}\mathbf{A}^*\right)_{j.} \left(\mathbf{a}_{i.}^*\right)\right) {}_\alpha^\alpha}{\sum\limits_{\alpha\in I_{r,\,m}} \left|\left(\mathbf{A}\mathbf{A}^*\right) {}_\alpha^\alpha\right|}. \tag{14}$$

Definition 2.7. *A Hermitian matrix* $\mathbf{A} \in \mathbb{H}^{n\times n}$ *is called positive (semi)definite if* $\mathbf{x}^*\mathbf{A}\mathbf{x} > 0 (\geq 0)$ *for any nonzero vector* $\mathbf{x} \in \mathbb{H}^n$.

For quaternion positive definite matrices, the following properties from the set of complex matrices can be obviously expanded. Their proofs are similar to the proofs in the complex case, then we list their without complete profs but with some comments.

Proposition 2.5. *Let* $\mathbf{A} \in \mathbb{H}^{n\times n}$ *be a positive definite matrix. Then following properties are equivalent.*

1. *All its eigenvalues are positive.*
2. *Its leading principal minors are all positive.*
3. *The associated sesquilinear form is an inner product.*
4. *It is the Gram matrix of linearly independent vectors.*
5. *It has a unique Cholesky decomposition.*

Proof.

1. By Corollary 2.4 there exists an unitary matrix $\mathbf{U}$ such that $\mathbf{A} = \mathbf{U}^*\mathbf{D}\mathbf{U}$, where right eigenvalues $\lambda_1, ..., \lambda_n$ are real. Then $\mathbf{x}^*\mathbf{A}\mathbf{x} = (\mathbf{U}\mathbf{x})^* \mathbf{D}\mathbf{U}\mathbf{x}$. By the one-to-one change of variable $\mathbf{y} = \mathbf{U}\mathbf{x}$, $\mathbf{y}^*\mathbf{A}\mathbf{y} > 0$, for any nonzero vector $\mathbf{y} \in \mathbb{H}^n$ when $\mathbf{D}$ is positive definite, It means each element of the main diagonal of $\mathbf{D}$ – that is, every eigenvalue of $\mathbf{A}$ – is positive.

2. Since all leading principal submatrices of a Hermitian matrix are Hermitian, then we can define leading principal minors as determinants of Hermitian submatrices in terms of (5).

3. Let the sesquilinear form by $\mathbf{A}$ be defined the function $\langle \cdot, \cdot \rangle_{r,A}$ from $\mathbb{H}^n \times \mathbb{H}^n$ to $\mathbb{H}$ such that $\langle \mathbf{x}, \mathbf{y} \rangle_{r,A} := \mathbf{y}^*\mathbf{A}\mathbf{x}$ for all $\mathbf{x}, \mathbf{y} \in \mathbb{H}^n$. By (2.1), the form is an inner product on $\mathbb{H}$ if and only if $\langle \mathbf{x}, \mathbf{x} \rangle_{r,A}$ is real and positive for all nonzero $\mathbf{x}$; that is if and only if $\mathbf{A}$ is positive definite.

4. Let $\mathbf{x}_1, \ldots, \mathbf{x}_n$ be a set of n linearly independent vectors of $\mathcal{H}_r^n$ with an inner product $\langle \cdot, \cdot \rangle_r$. It can be verified that the Gram matrix $\mathbf{A}$ of those entries, defined by $a_{ij} = \langle x_i, x_j \rangle_r$, is always positive definite.

5. There exists a unique lower triangular matrix $\mathbf{L}$, with real and strictly positive diagonal elements, such that $\mathbf{A} = \mathbf{L}\mathbf{L}^*$. This factorization is called the Cholesky decomposition of $\mathbf{A}$.

Every positive definite matrix $\mathbf{A} \in \mathbb{H}^{n\times n}$ has a unique square root defined by $\mathbf{A}^{\frac{1}{2}}$. It means, if $\mathbf{A} = \mathbf{U}\mathbf{D}\mathbf{U}^*$ then $\mathbf{A}^{\frac{1}{2}} = \mathbf{U}\mathbf{D}^{\frac{1}{2}}\mathbf{U}^*$.

Lemma 2.3. *[14] Let $\mathbf{A} \in \mathbb{H}_r^{m\times n}$, then $\mathbf{A}^*\mathbf{A}$ and $\mathbf{A}\mathbf{A}^*$ are both positive (semi)definite, and r nonzero eigenvalues of $\mathbf{A}^*\mathbf{A}$ and $\mathbf{A}\mathbf{A}^*$ coincide.*

Proof. The proof of the second part immediately follows from the singular value decomposition of $\mathbf{A} \in \mathbb{H}_r^{m\times n}$.□

Definition 2.8. *A square matrix $\mathbf{Q} \in \mathbb{H}^{m\times m}$ is called $\mathbf{H}$-weighted unitary (unitary with weight $\mathbf{H}$) if $\mathbf{Q}^*\mathbf{H}\mathbf{Q} = \mathbf{I}_m$, where $\mathbf{I}_m$ is the identity matrix.*

The following well-known two facts (see, e.g. [39]) on positive definite and Hermitian matrices and their product obviously can be extended to quaternion matrices.

Lemma 2.4. *Let* $\mathbf{A} \in \mathbb{H}^{n\times n}$ *be positive definite and* $\mathbf{B} \in \mathbb{H}^{n\times n}$ *be Hermitian matrices, respectively. Then* $\mathbf{AB}$ *is a diagonalizable matrix, it's all eigenvalues are real, and* $In\mathbf{AB} = In\mathbf{A}$.

Lemma 2.5. *Let* $\mathbf{A} \in \mathbb{H}^{n\times n}$ *be positive definite and* $\mathbf{B} \in \mathbb{H}^{n\times n}$ *be Hermitian matrices, respectively. Then there exists nonsingular* $\mathbf{C} \in \mathbb{H}^{n\times n}$ *such that* $\mathbf{C}^*\mathbf{AC} = \mathbf{I}_n$, *and* $\mathbf{C}^*\mathbf{BC} = \mathbf{\Lambda}$, *where* $\mathbf{\Lambda}$ *is a diagonal matrix.*

3. Weighted Singular Value Decomposition and Representations of the Weighted Moore-Penrose Inverse of Quaternion Matrices

3.1. Representations of the Weighted Moore-Penrose Inverse of Quaternion Matrices by WSVD

Denote $\mathbf{A}^\sharp = \mathbf{N}^{-1}\mathbf{A}^*\mathbf{M}$. Now, we prove the following theorem about the weighted singular value decomposition (WSVD) of quaternion matrices. A similar theorem for real matrices was independently proved by Loan [11] using the Cholesky decomposition, and by Galba [12], where WSVD of $\mathbf{A} \in \mathbb{R}^{m\times n}$ with positive definite weights $\mathbf{B}$ and $\mathbf{C}$ has been described as $\mathbf{A} = \mathbf{UDV}^T\mathbf{C}$. Our method of proving different from analogous for real matrices in [11], and has more similar manner to [12].

Theorem 3.1. *Let* $\mathbf{A} \in \mathbb{H}_r^{m\times n}$, *and* $\mathbf{M}$ *and* $\mathbf{N}$ *be positive definite matrices of order* m *and* n, *respectively. Then there exist* $\mathbf{U} \in \mathbb{H}^{m\times m}$, $\mathbf{V} \in \mathbb{H}^{n\times n}$ *satisfying* $\mathbf{U}^*\mathbf{MU} = \mathbf{I}_m$ *and* $\mathbf{V}^*\mathbf{N}^{-1}\mathbf{V} = \mathbf{I}_n$ *such that*

$$\mathbf{A} = \mathbf{UDV}^*, \tag{15}$$

where $\mathbf{D} = \begin{pmatrix} \mathbf{\Sigma} & \mathbf{0} \\ \mathbf{0} & \mathbf{0} \end{pmatrix}$, $\mathbf{\Sigma} = diag(\sigma_1, \sigma_2, ..., \sigma_r)$, $\sigma_1 \geq \sigma_2 \geq ... \geq \sigma_r > 0$ *and* σ_i^2 *is the nonzero eigenvalues of* $\mathbf{A}^\sharp\mathbf{A}$ *or* $\mathbf{AA}^\sharp$, *which coincide.*

Proof. First, consider $\mathbf{A}^\sharp\mathbf{A} = \mathbf{N}^{-1}\mathbf{A}^*\mathbf{MA}$. Since $\mathbf{A}^*\mathbf{MA} = (\mathbf{M}^{\frac{1}{2}}\mathbf{A})^*\mathbf{AM}^{\frac{1}{2}}$, then, by Lemma 2.3, $\mathbf{A}^*\mathbf{MA}$ is Hermitian positive semidefinite, and by Lemma 2.4 all eigenvalues of $\mathbf{A}^\sharp\mathbf{A}$ are nonnegative. Denote them by σ_i^2, where $\sigma_1 \geq ... \geq \sigma_n \geq 0$.

Denote $\mathbf{L} = \mathbf{A}^{\sharp}\mathbf{A}$. Since $\mathbf{L}\mathbf{N}^{-1} = \mathbf{N}^{-1}\mathbf{A}^{*}\mathbf{M}\mathbf{A}\mathbf{N}^{-1}$ is Hermitian and there exists a nonsingular $\mathbf{V} \in \mathbb{H}^{n\times n}$ such that $\mathbf{V}^{*}\mathbf{N}^{-1}\mathbf{V} = \mathbf{I}_n$, then by Lemma 2.5,

$$\mathbf{V}^{*}\mathbf{L}\mathbf{N}^{-1}\mathbf{V} = \mathbf{\Lambda}, \tag{16}$$

where $\mathbf{V}$ is unitary with weight $\mathbf{N}^{-1}$, and $\mathbf{\Lambda}$ is a diagonal matrix.

It follows from $\mathbf{L} = \mathbf{N}^{-1}\mathbf{V}\mathbf{\Lambda}\mathbf{V}^{*} = (\mathbf{V}^{*})^{-1}\mathbf{\Lambda}\mathbf{V}^{*}$ that $\mathbf{\Lambda} \equiv \mathbf{\Sigma}_1^2$, where $\mathbf{\Sigma}_1^2$ is diagonal with eigenvalues of $\mathbf{A}^{\sharp}\mathbf{A}$ on the principal diagonal, $\sigma_{ii}^2 = \sigma_i^2$ for all $i = 1, ..., n$. Since $\operatorname{rank} \mathbf{L} = \operatorname{rank} \mathbf{A} = r$, then the number of nonzero diagonal elements of $\mathbf{\Sigma}_1^2$ is equal r. Also, we note that

$$\mathbf{V}^{*}\mathbf{L}\mathbf{N}^{-1}\mathbf{V} = \mathbf{V}^{*}\mathbf{N}^{-1}\mathbf{A}^{*}\mathbf{M}\mathbf{A}\mathbf{N}^{-1}\mathbf{V} = \\ \mathbf{V}^{-1}\mathbf{A}^{*}(\mathbf{U}^{*})^{-1}\mathbf{U}^{-1}\mathbf{A}(\mathbf{V}^{*})^{-1} = \mathbf{\Sigma}_1^2. \tag{17}$$

Consider the following matrix,

$$\mathbf{P} = \mathbf{M}^{\frac{1}{2}}\mathbf{A}\mathbf{N}^{-1}\mathbf{V} \in \mathbb{H}^{m\times n}. \tag{18}$$

By virtue of (16),

$$\mathbf{P}^{*}\mathbf{P} = \left(\mathbf{V}^{*}\mathbf{N}^{-1}\mathbf{A}^{*}\mathbf{M}^{\frac{1}{2}}\right)\mathbf{M}^{\frac{1}{2}}\mathbf{A}\mathbf{N}^{-1}\mathbf{V} = \mathbf{\Sigma}_1^2. \tag{19}$$

Let us introduce the following $m \times n$ matrix $\mathbf{D} \in \mathbb{H}^{m\times n}$,

$$\mathbf{D} = \begin{pmatrix} \mathbf{\Sigma} & \mathbf{0} \\ \mathbf{0} & \mathbf{0} \end{pmatrix}, \tag{20}$$

where $\mathbf{\Sigma} \in \mathbb{H}^{r\times r}$ is a diagonal matrix with $\sigma_1 \geq \sigma_2 \geq ... \geq \sigma_r > 0$ on the principal diagonal. Then,

$$\mathbf{P} = \mathbf{M}^{\frac{1}{2}}\mathbf{U}\mathbf{D}. \tag{21}$$

By (18) and (21), we have $\mathbf{M}^{\frac{1}{2}}\mathbf{A}\mathbf{N}^{-1}\mathbf{V} = \mathbf{M}^{\frac{1}{2}}\mathbf{U}\mathbf{D}$. Due to the equality $\left(\mathbf{N}^{-1}\mathbf{V}\right)^{-1} = \mathbf{V}^{*}$, it follows (15).

Now we shall prove (15), where σ_i^2 is the nonzero eigenvalues of $\mathbf{A}\mathbf{A}^{\sharp} = \mathbf{A}\mathbf{N}^{-1}\mathbf{A}^{*}\mathbf{M}$. Since $\mathbf{A}\mathbf{N}^{-1}\mathbf{A}^{*}$ and $\mathbf{M}$ are respectively Hermitian positive semidefinite and definite, then by by Lemma 2.4 all eigenvalues of $\mathbf{A}\mathbf{A}^{\sharp}$ are nonnegative. Primarily, denote them by τ_i^2, where $\tau_1 \geq ... \geq \tau_m \geq 0$, and denote $\mathbf{Q} = \mathbf{A}\mathbf{A}^{\sharp}$. Since $\mathbf{M}\mathbf{Q} = \mathbf{M}\mathbf{A}\mathbf{N}^{-1}\mathbf{A}^{*}\mathbf{M}$ is Hermitian and there exists a nonsingular $\mathbf{U} \in \mathbb{H}^{m\times m}$ such that $\mathbf{U}^{*}\mathbf{M}\mathbf{U} = \mathbf{I}_m$, then by Lemma 2.5,

$$\mathbf{U}^{*}\mathbf{M}\mathbf{Q}\mathbf{U} = \mathbf{\Omega}, \tag{22}$$

where $\mathbf{U}$ is unitary with weight $\mathbf{M}$, and $\mathbf{\Omega}$ is a diagonal matrix.

It follows from $\mathbf{Q} = \mathbf{U\Omega U^*M} = \mathbf{U\Omega U}^{-1}$ that $\mathbf{\Omega} \equiv \mathbf{\Sigma}_2^2$, where $\mathbf{\Sigma}_2^2$ is diagonal with eigenvalues of $\mathbf{AA}^\sharp$ on the principal diagonal, $\tau_{ii}^2 = \tau_i^2$ for all $i = 1, ..., m$. Since rank $\mathbf{Q}$ = rank $\mathbf{A} = r$, then the number of nonzero diagonal elements of $\mathbf{\Sigma}_2^2$ is equal r. Also, we have

$$\mathbf{U^*MQU} = \mathbf{U^*MAN^{-1}A^*MU} = \mathbf{U}^{-1}\mathbf{A}\left(\mathbf{V}^*\right)^{-1}\mathbf{V}^{-1}\mathbf{A}^*\left(\mathbf{U}^*\right)^{-1} = \mathbf{\Sigma}_2^2. \quad (23)$$

Comparing (17) and (23), and due to Lemma 2.3, we have that r nonzero eigenvalues of $\mathbf{AA}^\sharp$ coincide with r nonzero eigenvalues of $\mathbf{A}^\sharp\mathbf{A}$, i.e. $\sigma_i^2 = \tau_i^2$ for all $i = 1, ..., r$.

Consider the following matrix,

$$\mathbf{S} = \mathbf{U^*MAN}^{-\frac{1}{2}} \in \mathbb{H}^{m\times n}. \quad (24)$$

By virtue of (22),

$$\mathbf{SS}^* = \mathbf{U^*MAN}^{-\frac{1}{2}}\left(\mathbf{N}^{-\frac{1}{2}}\mathbf{A^*MU}\right) = \mathbf{\Sigma}_2^2. \quad (25)$$

Consider again the matrix $\mathbf{D} \in \mathbb{H}^{m\times n}$ from (20). Then,

$$\mathbf{S} = \mathbf{DV^*N}^{-\frac{1}{2}}. \quad (26)$$

By (24) and (26), we have $\mathbf{U^*MAN}^{-\frac{1}{2}} = \mathbf{DV^*N}^{-\frac{1}{2}}$. From this, due to $\left(\mathbf{U^*M}\right)^{-1} = \mathbf{U}$, we again obtain (15).□

Now, we prove the following theorem about a representation of $\mathbf{A}^\dagger_{M,N}$ by WSVD of $\mathbf{A} \in \mathbb{H}_r^{m\times n}$ with weights $\mathbf{M}$ and $\mathbf{N}$.

Theorem 3.2. *Let* $\mathbf{A} \in \mathbb{H}_r^{m\times n}$, $\mathbf{M}$ *and* $\mathbf{N}$ *be positive definite matrices of order* m *and* n, *respectively. There exist* $\mathbf{U} \in \mathbb{H}^{m\times m}$, $\mathbf{V} \in \mathbb{H}^{n\times n}$ *satisfying* $\mathbf{U^*MU} = \mathbf{I}_m$ *and* $\mathbf{V^*N^{-1}V} = \mathbf{I}_n$ *such that* $\mathbf{A} = \mathbf{UDV}^*$, *where* $\mathbf{D} = \begin{pmatrix} \mathbf{\Sigma} & \mathbf{0} \\ \mathbf{0} & \mathbf{0} \end{pmatrix}$. *Then the weighted Moore-Penrose inverse* $\mathbf{A}^\dagger_{M,N}$ *can be represented*

$$\mathbf{A}^\dagger_{M,N} = \mathbf{N}^{-1}\mathbf{V}\begin{pmatrix} \mathbf{\Sigma}^{-1} & \mathbf{0} \\ \mathbf{0} & \mathbf{0} \end{pmatrix}\mathbf{U^*M}, \quad (27)$$

where $\mathbf{\Sigma} = diag(\sigma_1, \sigma_2, ..., \sigma_r)$, $\sigma_1 \geq \sigma_2 \geq ... \geq \sigma_r > 0$ *and* σ_i^2 *is the nonzero eigenvalues of* $\mathbf{A}^\sharp\mathbf{A}$ *or* $\mathbf{AA}^\sharp$, *which coincide.*

Proof. To prove the theorem it is enough to show that $\mathbf{X} = \mathbf{A}^{\dagger}_{M,N}$ expressed by (27) satisfies the equations (1), (2), (3N), and (4M).

$$1)\ \mathbf{AXA} = \mathbf{U}\begin{pmatrix} \boldsymbol{\Sigma} & \mathbf{0} \\ \mathbf{0} & \mathbf{0} \end{pmatrix}\mathbf{V}^*\mathbf{N}^{-1}\mathbf{V}\begin{pmatrix} \boldsymbol{\Sigma}^{-1} & \mathbf{0} \\ \mathbf{0} & \mathbf{0} \end{pmatrix}\mathbf{U}^*\mathbf{M}\mathbf{U}\begin{pmatrix} \boldsymbol{\Sigma} & \mathbf{0} \\ \mathbf{0} & \mathbf{0} \end{pmatrix}\mathbf{V}^* = \mathbf{A},$$

$$2)\ \mathbf{XAX} = \mathbf{N}^{-1}\mathbf{V}\begin{pmatrix} \boldsymbol{\Sigma}^{-1} & \mathbf{0} \\ \mathbf{0} & \mathbf{0} \end{pmatrix}\mathbf{U}^*\mathbf{M}\mathbf{U}\begin{pmatrix} \boldsymbol{\Sigma} & \mathbf{0} \\ \mathbf{0} & \mathbf{0} \end{pmatrix}\mathbf{V}^*\times$$
$$\mathbf{N}^{-1}\mathbf{V}\begin{pmatrix} \boldsymbol{\Sigma}^{-1} & \mathbf{0} \\ \mathbf{0} & \mathbf{0} \end{pmatrix}\mathbf{U}^*\mathbf{M} = \mathbf{X},$$

$$(3M)\ (\mathbf{MAX})^* = \left(\mathbf{MU}\begin{pmatrix} \boldsymbol{\Sigma} & \mathbf{0} \\ \mathbf{0} & \mathbf{0} \end{pmatrix}\mathbf{V}^*\mathbf{N}^{-1}\mathbf{V}\begin{pmatrix} \boldsymbol{\Sigma}^{-1} & \mathbf{0} \\ \mathbf{0} & \mathbf{0} \end{pmatrix}\mathbf{U}^*\mathbf{M}\right)^* =$$
$$\mathbf{MU}\begin{pmatrix} \mathbf{I} & \mathbf{0} \\ \mathbf{0} & \mathbf{0} \end{pmatrix}_{m\times m}\mathbf{U}^*\mathbf{M} = \mathbf{MU}\begin{pmatrix} \boldsymbol{\Sigma} & \mathbf{0} \\ \mathbf{0} & \mathbf{0} \end{pmatrix}\mathbf{V}^*\mathbf{N}^{-1}\mathbf{V}\begin{pmatrix} \boldsymbol{\Sigma}^{-1} & \mathbf{0} \\ \mathbf{0} & \mathbf{0} \end{pmatrix}\mathbf{U}^*\mathbf{M} =$$
$$\mathbf{MAX},$$

$$(4N)\ (\mathbf{NXA})^* = \left(\mathbf{NN}^{-1}\mathbf{V}\begin{pmatrix} \boldsymbol{\Sigma}^{-1} & \mathbf{0} \\ \mathbf{0} & \mathbf{0} \end{pmatrix}\mathbf{U}^*\mathbf{MU}\begin{pmatrix} \boldsymbol{\Sigma} & \mathbf{0} \\ \mathbf{0} & \mathbf{0} \end{pmatrix}\mathbf{V}^*\right)^* =$$
$$\mathbf{NN}^{-1}\mathbf{V}\begin{pmatrix} \mathbf{I} & \mathbf{0} \\ \mathbf{0} & \mathbf{0} \end{pmatrix}_{n\times n}\mathbf{V}^* = \mathbf{NN}^{-1}\mathbf{V}\begin{pmatrix} \boldsymbol{\Sigma}^{-1} & \mathbf{0} \\ \mathbf{0} & \mathbf{0} \end{pmatrix}\mathbf{U}^*\mathbf{MU}\begin{pmatrix} \boldsymbol{\Sigma} & \mathbf{0} \\ \mathbf{0} & \mathbf{0} \end{pmatrix}\mathbf{V}^* =$$
$$\mathbf{NXA}.$$

□

3.2. Limit Representations of the Weighted Moore-Penrose Inverse Over the Quaternion Skew Field

Due to [3] the following limit representation can be extended to $\mathbb{H}$. We give the proof of the following lemma that different from ([3], Corollary 3.4.) and based on WSVD.

Lemma 3.1. *Let $\mathbf{A} \in \mathbb{H}_r^{m\times n}$, and $\mathbf{M}$ and $\mathbf{N}$ be positive definite matrices of order m and n, respectively. Then*

$$\mathbf{A}^{\dagger}_{M,N} = \lim_{\lambda\to 0}(\lambda\mathbf{I} + \mathbf{A}^{\sharp}\mathbf{A})^{-1}\mathbf{A}^{\sharp}. \tag{28}$$

where $\mathbf{A}^{\sharp} = \mathbf{N}^{-1}\mathbf{A}^*\mathbf{M}$, $\lambda \in \mathbb{R}_+$ and $\mathbb{R}_+$ is the set of all positive real numbers.

Proof. By Theorems 3.1 and 3.2, respectively, we have

$$\mathbf{A} = \mathbf{U}\begin{pmatrix} \mathbf{\Sigma} & \mathbf{0} \\ \mathbf{0} & \mathbf{0} \end{pmatrix}\mathbf{V}^*, \ \mathbf{A}^{\dagger}_{M,N} = \mathbf{N}^{-1}\mathbf{V}\begin{pmatrix} \mathbf{\Sigma}^{-1} & \mathbf{0} \\ \mathbf{0} & \mathbf{0} \end{pmatrix}\mathbf{U}^*\mathbf{M},$$

where $\mathbf{\Sigma} = diag(\sigma_1, \sigma_2, ..., \sigma_r)$, $\sigma_1 \geq \sigma_2 \geq ... \geq \sigma_r > 0$ and $\sigma_i^2 \in \mathbb{R}$ is the nonzero eigenvalues of $\mathbf{N}^{-1}\mathbf{A}^*\mathbf{M}\mathbf{A}$. Consider the matrix

$$\mathbf{D} := \begin{pmatrix} \mathbf{\Sigma} & \mathbf{0} \\ \mathbf{0} & \mathbf{0} \end{pmatrix},$$

where $\mathbf{D} = (\sigma_{ij}) \in \mathbb{H}^{m\times n}_r$ is such that $\sigma_{11} \geq \sigma_{22} \geq \ldots \geq \sigma_{rr} > \sigma_{r+1\,r+1} = \ldots = \sigma_{qq} = 0$, $q = \min\{n, m\}$. Then

$$\mathbf{D}^* = \begin{pmatrix} \mathbf{\Sigma}^* & \mathbf{0} \\ \mathbf{0} & \mathbf{0} \end{pmatrix}, \ \mathbf{D}^{\dagger} = \begin{pmatrix} \mathbf{\Sigma}^{-1} & \mathbf{0} \\ \mathbf{0} & \mathbf{0} \end{pmatrix},$$

and $\mathbf{A} = \mathbf{U}\mathbf{D}\mathbf{V}^*$, $\mathbf{A}^{\sharp} = \mathbf{N}^{-1}\mathbf{V}\mathbf{D}^*\mathbf{U}^*\mathbf{M}$, $\mathbf{A}^{\dagger}_{M,N} = \mathbf{N}^{-1}\mathbf{V}\mathbf{D}^{\dagger}\mathbf{U}^*\mathbf{M}$. Since $\mathbf{N}^{-1}\mathbf{V} = (\mathbf{V}^*)^{-1}$, then we have

$$\lambda\mathbf{I} + \mathbf{A}^{\sharp}\mathbf{A} = \lambda\mathbf{I} + \mathbf{N}^{-1}\mathbf{V}\mathbf{D}^*\mathbf{U}^*\mathbf{M}\mathbf{U}\mathbf{D}\mathbf{V}^* = \lambda\mathbf{I} + (\mathbf{V}^*)^{-1}\mathbf{D}^*\mathbf{D}\mathbf{V}^* = (\mathbf{V}^*)^{-1}(\lambda\mathbf{I} + \mathbf{D}^2)\mathbf{V}^*.$$

Further,

$$(\lambda\mathbf{I} + \mathbf{A}^{\sharp}\mathbf{A})^{-1}\mathbf{A}^{\sharp} = (\mathbf{V}^*)^{-1}(\lambda\mathbf{I} + \mathbf{D}^2)^{-1}\mathbf{V}^*\mathbf{N}^{-1}\mathbf{V}\mathbf{D}^*\mathbf{U}^*\mathbf{M} = \mathbf{N}^{-1}\mathbf{V}(\lambda\mathbf{I} + \mathbf{D}^2)^{-1}\mathbf{D}^*\mathbf{U}^*\mathbf{M}.$$

Consider the matrix

$$\left(\lambda\mathbf{I} + \mathbf{D}^2\right)^{-1}\mathbf{D} = \begin{pmatrix} \frac{\sigma_1}{\sigma_1^2+\lambda} & \ldots & 0 & \\ \ldots & \ldots & \ldots & \mathbf{0} \\ 0 & \ldots & \frac{\sigma_r}{\sigma_r^2+\lambda} & \vdots \\ & \vdots & & \ddots \\ & \mathbf{0} & & \mathbf{0} \end{pmatrix}.$$

It is obviously that $\lim\limits_{\lambda\to 0}\left(\lambda\mathbf{I} + \mathbf{D}^2\right)^{-1}\mathbf{D} = \mathbf{D}^{\dagger}$. Then,

$$\lim_{\lambda\to 0}\mathbf{N}^{-1}\mathbf{V}(\lambda\mathbf{I} + \mathbf{D}^2)^{-1}\mathbf{D}^*\mathbf{U}^*\mathbf{M} = \mathbf{N}^{-1}\mathbf{V}\mathbf{D}^{\dagger}\mathbf{U}^*\mathbf{M} = \mathbf{A}^{\dagger}_{M,N}.$$

The lemma is proofed.□

The following lemma gives another limit representation of $\mathbf{A}^{\dagger}_{M,N}$.

Lemma 3.2. *Let* $\mathbf{A} \in \mathbb{H}^{m\times n}_{r}$, *and* $\mathbf{M}$ *and* $\mathbf{N}$ *be positive definite matrices of order* m *and* n, *respectively. Then*

$$\mathbf{A}^{\dagger}_{M,N} = \lim_{\lambda\to 0} \mathbf{A}^{\sharp}(\lambda\mathbf{I} + \mathbf{A}\mathbf{A}^{\sharp})^{-1},$$

where $\mathbf{A}^{\sharp} = \mathbf{N}^{-1}\mathbf{A}^{*}\mathbf{M}$, $\lambda \in \mathbb{R}_{+}$.

Proof. The proof is similar to the proof of Lemma 3.1 by using the fact from Theorem 3.1 that the nonzero eigenvalues of $\mathbf{A}^{\sharp}\mathbf{A}$ and $\mathbf{A}\mathbf{A}^{\sharp}$ coincide.□

It is evidently the following corollary.

Corollary 3.1. *If* $\mathbf{A} \in \mathbb{H}^{m\times n}$, *then the following statements are true.*

i) If $\operatorname{rank}\mathbf{A} = \mathrm{n}$, *then* $\mathbf{A}^{\dagger}_{M,N} = \left(\mathbf{A}^{\sharp}\mathbf{A}\right)^{-1}\mathbf{A}^{\sharp}$.

ii) If $\operatorname{rank}\mathbf{A} = \mathrm{m}$, *then* $\mathbf{A}^{\dagger}_{M,N} = \mathbf{A}^{\sharp}\left(\mathbf{A}\mathbf{A}^{\sharp}\right)^{-1}$.

iii) If $\operatorname{rank}\mathbf{A} = \mathrm{n} = \mathrm{m}$, *then* $\mathbf{A}^{\dagger}_{M,N} = \mathbf{A}^{-1}$.

4. Determinantal Representations of the Weighted Moore-Penrose Inverse Over the Quaternion Skew Field

Even though the eigenvalues of $\mathbf{A}^{\sharp}\mathbf{A}$ and $\mathbf{A}\mathbf{A}^{\sharp}$ are real and nonnegative, they are not Hermitian in general. Therefor, we consider two cases, when $\mathbf{A}^{\sharp}\mathbf{A}$ and $\mathbf{A}\mathbf{A}^{\sharp}$ both or one of them are Hermitian, and when they are non-Hermitian.

4.1. The Case of Hermitian $\mathbf{A}^{\sharp}\mathbf{A}$ and $\mathbf{A}\mathbf{A}^{\sharp}$

Let $(\mathbf{A}^{\sharp}\mathbf{A}) \in \mathbb{H}^{n\times n}$ be Hermitian. It means that $(\mathbf{A}^{\sharp}\mathbf{A})^{*} = (\mathbf{A}^{\sharp}\mathbf{A})$. Since $\mathbf{N}^{-1}$ and $\mathbf{M}$ are Hermitian, then

$$(\mathbf{N}^{-1}\mathbf{A}^{*}\mathbf{M}\mathbf{A})^{*} = \mathbf{A}^{*}\mathbf{M}\mathbf{A}\mathbf{N}^{-1} = \mathbf{N}^{-1}\mathbf{A}^{*}\mathbf{M}\mathbf{A}.$$

So, to the matrix $(\mathbf{A}^{\sharp}\mathbf{A})$ be Hermitian the matrices $\mathbf{N}^{-1}$ and $(\mathbf{A}^{*}\mathbf{M}\mathbf{A})$ should be commutative. Similarly, to $(\mathbf{A}\mathbf{A}^{\sharp})$ be Hermitian the matrices $\mathbf{M}$ and $(\mathbf{A}\mathbf{N}^{-1}\mathbf{A}^{*})$ should be commutative.

Denote by $\mathbf{a}^{\sharp}_{.j}$ and $\mathbf{a}^{\sharp}_{i.}$ the jth column and the ith row of $\mathbf{A}^{\sharp}$ respectively.

Lemma 4.1. *If* $\mathbf{A} \in \mathbb{H}_r^{m\times n}$, *then* $\operatorname{rank}\left(\mathbf{A}^\sharp \mathbf{A}\right)_{.\,i}\left(\mathbf{a}^\sharp_{.j}\right) \leq r$.

Proof. Let's lead elementary transformations of the matrix $\left(\mathbf{A}^\sharp \mathbf{A}\right)_{.\,i}\left(\mathbf{a}^\sharp_{.j}\right)$ right-multiplying it by elementary unimodular matrices $\mathbf{P}_{i\,k}\left(-a_{jk}\right)$, $k \neq j$. The matrix $\mathbf{P}_{i\,k}\left(-a_{jk}\right)$ has $-a_{j\,k}$ in the (i,k) entry, 1 in all diagonal entries, and 0 in others. It is the matrix of an elementary transformation. Right-multiplying a matrix by $\mathbf{P}_{i\,k}\left(-a_{jk}\right)$, where $k \neq j$, means adding to k-th column its i-th column right-multiplying on $-a_{jk}$. Then we get

$$\left(\mathbf{A}^\sharp \mathbf{A}\right)_{.\,i}\left(\mathbf{a}^\sharp_{.\,j}\right)\cdot\prod_{k\neq i}\mathbf{P}_{i\,k}\left(-a_{j\,k}\right)=\begin{pmatrix} \sum\limits_{k\neq j} a^\sharp_{1k}a_{k1} & \dots & a^\sharp_{1j} & \dots & \sum\limits_{k\neq j} a^\sharp_{1k}a_{kn} \\ \dots & \dots & \dots & \dots & \dots \\ \sum\limits_{k\neq j} a^\sharp_{nk}a_{k1} & \dots & a^\sharp_{nj} & \dots & \sum\limits_{k\neq j} a^\sharp_{nk}a_{kn} \end{pmatrix}.$$
$$\hphantom{xxxxxxxxxxxxxxxx} ith$$

The obtained matrix has the following factorization.

$$\begin{pmatrix} \sum\limits_{k\neq j} a^\sharp_{1k}a_{k1} & \dots & a^\sharp_{1j} & \dots & \sum\limits_{k\neq j} a^\sharp_{1k}a_{kn} \\ \dots & \dots & \dots & \dots & \dots \\ \sum\limits_{k\neq j} a^\sharp_{nk}a_{k1} & \dots & a^\sharp_{nj} & \dots & \sum\limits_{k\neq j} a^\sharp_{nk}a_{kn} \end{pmatrix} =$$
$$ith$$

$$= \begin{pmatrix} a^\sharp_{11} & a^\sharp_{12} & \dots & a^\sharp_{1m} \\ a^\sharp_{21} & a^\sharp_{22} & \dots & a^\sharp_{2m} \\ \dots & \dots & \dots & \dots \\ a^\sharp_{n1} & a^\sharp_{n2} & \dots & a^\sharp_{nm} \end{pmatrix}\begin{pmatrix} a_{11} & \dots & 0 & \dots & a_{n1} \\ \dots & \dots & \dots & \dots & \dots \\ 0 & \dots & 1 & \dots & 0 \\ \dots & \dots & \dots & \dots & \dots \\ a_{m1} & \dots & 0 & \dots & a_{mn} \end{pmatrix} jth.$$
$$ith$$

Denote by $\tilde{\mathbf{A}} := \begin{pmatrix} a_{11} & \dots & 0 & \dots & a_{n1} \\ \dots & \dots & \dots & \dots & \dots \\ 0 & \dots & 1 & \dots & 0 \\ \dots & \dots & \dots & \dots & \dots \\ a_{m1} & \dots & 0 & \dots & a_{mn} \end{pmatrix} jth$ (ith column). The matrix $\tilde{\mathbf{A}}$ is obtained from $\mathbf{A}$ by replacing all entries of the jth row and of the ith column with zeroes except that the (j,i)th entry equals 1. Elementary transformations of a matrix do not change its rank and a rank of a matrix product does not exceed ranks of factors. It follows that $\operatorname{rank}\left(\mathbf{A}^\sharp \mathbf{A}\right)_{.\,i}\left(\mathbf{a}^\sharp_{.j}\right) \leq$

$\min\left\{\operatorname{rank}\mathbf{A}^{\sharp}, \operatorname{rank}\tilde{\mathbf{A}}\right\}$. It is obviously that $\operatorname{rank}\tilde{\mathbf{A}} \geq \operatorname{rank}\mathbf{A} = \operatorname{rank}\mathbf{A}^{\sharp}$. This completes the proof. □

The following lemma can be proved similarly.

Lemma 4.2. *If* $\mathbf{A} \in \mathbb{H}_r^{m\times n}$, *then* $\operatorname{rank}\left(\mathbf{A}\mathbf{A}^{\sharp}\right)_{.\,i}\left(\mathbf{a}^{\sharp}_{.j}\right) \leq r$.

Analogues of the characteristic polynomial are considered in the following lemmas.

Lemma 4.3. *If* $\mathbf{A} \in \mathbb{H}^{m\times n}$, $t \in \mathbb{R}$, *and* $(\mathbf{A}^{\sharp}\mathbf{A})$ *is Hermitian, then*

$$\operatorname{cdet}_i\left(t\mathbf{I} + \mathbf{A}^{\sharp}\mathbf{A}\right)_{.\,i}\left(\mathbf{a}^{\sharp}_{.j}\right) = c_1^{(ij)}t^{n-1} + c_2^{(ij)}t^{n-2} + \cdots + c_n^{(ij)}, \tag{29}$$

where $c_n^{(ij)} = \operatorname{cdet}_i\left(\mathbf{A}^{\sharp}\mathbf{A}\right)_{.\,i}\left(\mathbf{a}^{\sharp}_{.j}\right)$ *and*

$$c_k^{(ij)} = \sum_{\beta\in J_{k,\,n}\{i\}} \operatorname{cdet}_i\left(\left(\mathbf{A}^{\sharp}\mathbf{A}\right)_{.\,i}\left(\mathbf{a}^{\sharp}_{.\,j}\right)\right)_{\beta}^{\beta}$$

for all $k = 1,\ldots,n-1$, $i = 1,\ldots,n$, *and* $j = 1,\ldots,m$.

Proof. Denote by $\mathbf{b}_{.i}$ the ith column of the Hermitian matrix $\mathbf{A}^{\sharp}\mathbf{A} =: (b_{ij})_{n\times n}$. Consider the Hermitian matrix $\left(t\mathbf{I} + \mathbf{A}^{\sharp}\mathbf{A}\right)_{.i}(\mathbf{b}_{.i}) \in \mathbb{H}^{n\times n}$. It differs from $\left(t\mathbf{I} + \mathbf{A}^{\sharp}\mathbf{A}\right)$ in an entry b_{ii}. Taking into account Lemma 2.1, we obtain

$$\det\left(t\mathbf{I} + \mathbf{A}^{\sharp}\mathbf{A}\right)_{.\,i}(\mathbf{b}_{.i}) = d_1t^{n-1} + d_2t^{n-2} + \ldots + d_n, \tag{30}$$

where $d_k = \sum\limits_{\beta\in J_{k,\,n}\{i\}} \det\left(\mathbf{A}^{\sharp}\mathbf{A}\right)_{\beta}^{\beta}$ is the sum of all principal minors of order k that contain the ith column for all $k = 1,\ldots,n-1$ and $d_n = \det\left(\mathbf{A}^{\sharp}\mathbf{A}\right)$. Therefore, we have

$$\mathbf{b}_{.\,i} = \begin{pmatrix} \sum\limits_l a^{\sharp}_{1l}a_{li} \\ \sum\limits_l a^{\sharp}_{2l}a_{li} \\ \vdots \\ \sum\limits_l a^{\sharp}_{nl}a_{li} \end{pmatrix} = \sum_l \mathbf{a}^{\sharp}_{.\,l}a_{li},$$

where $\mathbf{a}^{\sharp}_{.l}$ is the lth column-vector of $\mathbf{A}^{\sharp}$ for all $l = 1, \ldots, m$. Taking into account Theorem 2.1, Remark 2.1 and Proposition 2.1, on the one hand we obtain

$$\det\left(t\mathbf{I} + \mathbf{A}^{\sharp}\mathbf{A}\right)_{.i}(\mathbf{b}_{.i}) = \operatorname{cdet}_i\left(t\mathbf{I} + \mathbf{A}^{\sharp}\mathbf{A}\right)_{.i}(\mathbf{b}_{.i}) =$$
$$= \sum_l \operatorname{cdet}_i\left(t\mathbf{I} + \mathbf{A}^{\sharp}\mathbf{A}\right)_{.l}\left(\mathbf{a}^{\sharp}_{.l}a_{li}\right) = \sum_l \operatorname{cdet}_i\left(t\mathbf{I} + \mathbf{A}^{\sharp}\mathbf{A}\right)_{.i}\left(\mathbf{a}^{\sharp}_{.l}\right) \cdot a_{li} \tag{31}$$

On the other hand having changed the order of summation, we get for all $k = 1, \ldots, n-1$

$$d_k = \sum_{\beta \in J_{k,n}\{i\}} \det\left(\mathbf{A}^{\sharp}\mathbf{A}\right)^{\beta}_{\beta} = \sum_{\beta \in J_{k,n}\{i\}} \operatorname{cdet}_i\left(\mathbf{A}^{\sharp}\mathbf{A}\right)^{\beta}_{\beta} =$$
$$\sum_{\beta \in J_{k,n}\{i\}} \sum_l \operatorname{cdet}_i\left(\left(\mathbf{A}^{\sharp}\mathbf{A}\right)_{.i}\left(\mathbf{a}^{\sharp}_{.l}a_{li}\right)\right)^{\beta}_{\beta} =$$
$$\sum_l \sum_{\beta \in J_{k,n}\{i\}} \operatorname{cdet}_i\left(\left(\mathbf{A}^{\sharp}\mathbf{A}\right)_{.i}\left(\mathbf{a}^{\sharp}_{.l}\right)\right)^{\beta}_{\beta} \cdot a_{li}. \tag{32}$$

By substituting (31) and (32) in (30), and equating factors at a_{li} when $l = j$, we obtain the equality (29). □

The following lemma can be proved similarly.

Lemma 4.4. *If* $\mathbf{A} \in \mathbb{H}^{m\times n}$ *and* $t \in \mathbb{R}$, *and* $\mathbf{A}\mathbf{A}^{\sharp}$ *is Hermitian, then*

$$\operatorname{rdet}_j(t\mathbf{I} + \mathbf{A}\mathbf{A}^{\sharp})_{j.}(\mathbf{a}^{\sharp}_{i.}) = r_1^{(ij)}t^{n-1} + r_2^{(ij)}t^{n-2} + \cdots + r_n^{(ij)},$$

where $r_n^{(ij)} = \operatorname{rdet}_j(\mathbf{A}\mathbf{A}^{\sharp})_{j.}(\mathbf{a}^{\sharp}_{i.})$ *and*

$$r_k^{(ij)} = \sum_{\alpha \in I_{r,m}\{j\}} \operatorname{rdet}_j\left((\mathbf{A}\mathbf{A}^{\sharp})_{j.}(\mathbf{a}^{\sharp}_{i.})\right)^{\alpha}_{\alpha}$$

for all $k = 1, \ldots, n-1$, $i = 1, \ldots, n$, *and* $j = 1, \ldots, m$.

The following theorem introduces the determinantal representations of the weighted Moore-Penrose by analogs of the classical adjoint matrix.

Denote the (ij)th entry of $\mathbf{A}^{\dagger}_{M,N}$ by $a^{\ddagger}_{ij}$ for all $i = 1, \ldots, n$ and $j = 1, \ldots, m$.

Theorem 4.1. *Let* $\mathbf{A} \in \mathbb{H}_r^{m\times n}$. *If* $\mathbf{A}^\sharp\mathbf{A}$ *or* $\mathbf{A}\mathbf{A}^\sharp$ *are Hermitian, then the weighted Moore-Penrose inverse* $\mathbf{A}_{M,N}^\dagger = \left(a_{ij}^\ddagger\right) \in \mathbb{H}^{n\times m}$ *possess the following determinantal representations, respectively,*

$$a_{ij}^\ddagger = \frac{\sum\limits_{\beta\in J_{r,\,n}\{i\}} \operatorname{cdet}_i\left(\left(\mathbf{A}^\sharp\mathbf{A}\right)_{.\,i}\left(\mathbf{a}_{.j}^\sharp\right)\right)_\beta^\beta}{\sum\limits_{\beta\in J_{r,\,n}} \left|\left(\mathbf{A}^\sharp\mathbf{A}\right)_\beta^\beta\right|}, \tag{33}$$

or

$$a_{ij}^\ddagger = \frac{\sum\limits_{\alpha\in I_{r,m}\{j\}} \operatorname{rdet}_j\left((\mathbf{A}\mathbf{A}^\sharp)_{j\,.}(\mathbf{a}_{i.}^\sharp)\right)_\alpha^\alpha}{\sum\limits_{\alpha\in I_{r,\,m}} |(\mathbf{A}\mathbf{A}^\sharp)_\alpha^\alpha|}. \tag{34}$$

Proof. At first we prove (33). By Lemma 3.1, $\mathbf{A}^\dagger = \lim\limits_{\lambda\to 0}\left(\lambda\mathbf{I}+\mathbf{A}^\sharp\mathbf{A}\right)^{-1}\mathbf{A}^\sharp$. Let $\mathbf{A}^\sharp\mathbf{A}$ is Hermitian. Then the matrix $\left(\lambda\mathbf{I}+\mathbf{A}^\sharp\mathbf{A}\right) \in \mathbb{H}^{n\times n}$ is a full-rank Hermitian matrix for $\forall\lambda \in \mathbb{R}_+$. Taking into account Theorem 2.4 it has an inverse, which we represent as a left inverse matrix

$$\left(\lambda\mathbf{I}+\mathbf{A}^\sharp\mathbf{A}\right)^{-1} = \frac{1}{\det\left(\lambda\mathbf{I}+\mathbf{A}^\sharp\mathbf{A}\right)}\begin{pmatrix} L_{11} & L_{21} & \dots & L_{n1}\\ L_{12} & L_{22} & \dots & L_{n2}\\ \dots & \dots & \dots & \dots\\ L_{1n} & L_{2n} & \dots & L_{nn}\end{pmatrix},$$

where L_{ij} is a left (ij)th cofactor of $\alpha\mathbf{I}+\mathbf{A}^\sharp\mathbf{A}$. Then we have

$$\left(\lambda\mathbf{I}+\mathbf{A}^\sharp\mathbf{A}\right)^{-1}\mathbf{A}^\sharp = $$
$$= \frac{1}{\det\left(\lambda\mathbf{I}+\mathbf{A}^\sharp\mathbf{A}\right)}\begin{pmatrix} \sum\limits_{k=1}^n L_{k1}a_{k1}^\sharp & \sum\limits_{k=1}^n L_{k1}a_{k2}^\sharp & \dots & \sum\limits_{k=1}^n L_{k1}a_{km}^\sharp\\ \sum\limits_{k=1}^n L_{k2}a_{k1}^\sharp & \sum\limits_{k=1}^n L_{k2}a_{k2}^\sharp & \dots & \sum\limits_{k=1}^n L_{k2}a_{km}^\sharp\\ \dots & \dots & \dots & \dots\\ \sum\limits_{k=1}^n L_{kn}a_{k1}^\sharp & \sum\limits_{k=1}^n L_{kn}a_{k2}^\sharp & \dots & \sum\limits_{k=1}^n L_{kn}a_{km}^\sharp\end{pmatrix}.$$

Using the definition of a left cofactor, we obtain

$$\mathbf{A}^{\dagger}_{M,N} = \lim_{\alpha\to 0} \begin{pmatrix} \frac{\mathrm{cdet}_1\left(\lambda\mathbf{I}+\mathbf{A}^{\sharp}\mathbf{A}\right)_{.1}\left(\mathbf{a}^{\sharp}_{.1}\right)}{\det\left(\lambda\mathbf{I}+\mathbf{A}^{\sharp}\mathbf{A}\right)} & \cdots & \frac{\mathrm{cdet}_1\left(\lambda\mathbf{I}+\mathbf{A}^{\sharp}\mathbf{A}\right)_{.1}\left(\mathbf{a}^{\sharp}_{.m}\right)}{\det\left(\lambda\mathbf{I}+\mathbf{A}^{\sharp}\mathbf{A}\right)} \\ \cdots & \cdots & \cdots \\ \frac{\mathrm{cdet}_n\left(\lambda\mathbf{I}+\mathbf{A}^{\sharp}\mathbf{A}\right)_{.n}\left(\mathbf{a}^{\sharp}_{.1}\right)}{\det\left(\lambda\mathbf{I}+\mathbf{A}^{\sharp}\mathbf{A}\right)} & \cdots & \frac{\mathrm{cdet}_n\left(\lambda\mathbf{I}+\mathbf{A}^{\sharp}\mathbf{A}\right)_{.n}\left(\mathbf{a}^{\sharp}_{.m}\right)}{\det\left(\lambda\mathbf{I}+\mathbf{A}^{\sharp}\mathbf{A}\right)} \end{pmatrix}. \tag{35}$$

By Lemma 2.1, we have $\det\left(\lambda\mathbf{I}+\mathbf{A}^{\sharp}\mathbf{A}\right) = \lambda^n + d_1\lambda^{n-1} + d_2\lambda^{n-2} + \cdots + d_n$, where $d_k = \sum\limits_{\beta\in J_{k,\,n}} \left|\left(\mathbf{A}^{\sharp}\mathbf{A}\right)^{\beta}_{\beta}\right|$ is a sum of principal minors of $\mathbf{A}^{\sharp}\mathbf{A}$ of order k for all $k = 1,\ldots,n-1$ and $d_n = \det\mathbf{A}^{\sharp}\mathbf{A}$. Since $\mathrm{rank}\,\mathbf{A}^{\sharp}\mathbf{A} = \mathrm{rank}\,\mathbf{A} = r$ and $d_n = d_{n-1} = \cdots = d_{r+1} = 0$, it follows that

$$\det\left(\lambda\mathbf{I}+\mathbf{A}^{\sharp}\mathbf{A}\right) = \lambda^n + d_1\lambda^{n-1} + d_2\lambda^{n-2} + \ldots + d_r\lambda^{n-r}.$$

Using (29), we have

$$\mathrm{cdet}_i\left(\lambda\mathbf{I}+\mathbf{A}^{*}\mathbf{A}\right)_{.i}\left(\mathbf{a}^{\sharp}_{.j}\right) = c_1^{(ij)}\lambda^{n-1} + c_2^{(ij)}\lambda^{n-2} + \ldots + c_n^{(ij)}$$

for all $i = 1,\ldots,n$ and $j = 1,\ldots,m$, where

$$c_k^{(ij)} = \sum_{\beta\in J_{k,\,n}\{i\}} \mathrm{cdet}_i\left(\left(\mathbf{A}^{\sharp}\mathbf{A}\right)_{.i}\left(\mathbf{a}^{\sharp}_{.j}\right)\right)^{\beta}_{\beta}$$

for all $k = 1,\ldots,n-1$ and $c_n^{(ij)} = \mathrm{cdet}_i\left(\mathbf{A}^{\sharp}\mathbf{A}\right)_{.i}\left(\mathbf{a}^{\sharp}_{.j}\right)$.

Now we prove that $c_k^{(ij)} = 0$, when $k \geq r+1$ for all $i = 1,\ldots,n$, and $j = 1,\ldots,m$. By Lemma 4.1 $\mathrm{rank}\left(\mathbf{A}^{\sharp}\mathbf{A}\right)_{.i}\left(\mathbf{a}^{\sharp}_{.j}\right) \leq r$, then the matrix $\left(\mathbf{A}^{\sharp}\mathbf{A}\right)_{.i}\left(\mathbf{a}^{\sharp}_{.j}\right)$ has no more r right-linearly independent columns.

Consider $\left(\left(\mathbf{A}^{\sharp}\mathbf{A}\right)_{.i}\left(\mathbf{a}^{\sharp}_{.j}\right)\right)^{\beta}_{\beta}$, when $\beta\in J_{k,n}\{i\}$. It is a principal submatrix of $\left(\mathbf{A}^{\sharp}\mathbf{A}\right)_{.i}\left(\mathbf{a}^{\sharp}_{.j}\right)$ of order $k \geq r+1$. Deleting both its ith row and column, we obtain a principal submatrix of order $k-1$ of $\mathbf{A}^{\sharp}\mathbf{A}$. We denote it by $\mathbf{M}$. The following cases are possible.

Let $k = r+1$ and $\det\mathbf{M} \neq 0$. In this case all columns of $\mathbf{M}$ are right-linearly independent. The addition of all of them on one coordinate to columns of $\left(\left(\mathbf{A}^{\sharp}\mathbf{A}\right)_{.i}\left(\mathbf{a}^{\sharp}_{.j}\right)\right)^{\beta}_{\beta}$ keeps their right-linear independence. Hence, they are basis in a matrix $\left(\left(\mathbf{A}^{\sharp}\mathbf{A}\right)_{.i}\left(\mathbf{a}^{\sharp}_{.j}\right)\right)^{\beta}_{\beta}$, and the ith column is the right linear combination of its basic columns. From this by Theorem 2.5, we get $\mathrm{cdet}_i\left(\left(\mathbf{A}^{\sharp}\mathbf{A}\right)_{.i}\left(\mathbf{a}^{\sharp}_{.j}\right)\right)^{\beta}_{\beta} = 0$, when $\beta \in J_{k,n}\{i\}$ and $k \geq r+1$.

If $k = r+1$ and $\det \mathbf{M} = 0$, than p, ($p < k$), columns are basis in $\mathbf{M}$ and in $\left(\left(\mathbf{A}^{\sharp}\mathbf{A}\right)_{.i}\left(\mathbf{a}^{\sharp}_{.j}\right)\right)^{\beta}_{\beta}$. Then by Theorem 2.5, we obtain $\mathrm{cdet}_i\left(\left(\mathbf{A}^{\sharp}\mathbf{A}\right)_{.i}\left(\mathbf{a}^{\sharp}_{.j}\right)\right)^{\beta}_{\beta} = 0$ as well.

If $k > r+1$, then by Theorem 2.6 it follows that $\det \mathbf{M} = 0$ and p, ($p < k-1$), columns are basis in the both matrices $\mathbf{M}$ and $\left(\left(\mathbf{A}^{\sharp}\mathbf{A}\right)_{.i}\left(\mathbf{a}^{\sharp}_{.j}\right)\right)^{\beta}_{\beta}$. Therefore, by Theorem 2.5, we obtain $\mathrm{cdet}_i\left(\left(\mathbf{A}^{\sharp}\mathbf{A}\right)_{.i}\left(\mathbf{a}^{\sharp}_{.j}\right)\right)^{\beta}_{\beta} = 0$.

Thus in all cases, we have $\mathrm{cdet}_i\left(\left(\mathbf{A}^{\sharp}\mathbf{A}\right)_{.i}\left(\mathbf{a}^{\sharp}_{.j}\right)\right)^{\beta}_{\beta} = 0$, when $\beta \in J_{k,n}\{i\}$ and $r+1 \leq k < n$, and for all $i = 1, \ldots, n$ and $j = 1, \ldots, m$,

$$c_k^{(ij)} = \sum_{\beta \in J_{k,\,n}\{i\}} \mathrm{cdet}_i\left(\left(\mathbf{A}^{\sharp}\mathbf{A}\right)_{.i}\left(\mathbf{a}^{\sharp}_{.j}\right)\right)^{\beta}_{\beta} = 0,$$

$$c_n^{(ij)} = \mathrm{cdet}_i\left(\mathbf{A}^{\sharp}\mathbf{A}\right)_{.i}\left(\mathbf{a}^{\sharp}_{.j}\right) = 0.$$

Hence, $\mathrm{cdet}_i\left(\lambda\mathbf{I} + \mathbf{A}^{\sharp}\mathbf{A}\right)_{.i}\left(\mathbf{a}^{\sharp}_{.j}\right) = c_1^{(ij)}\lambda^{n-1} + c_2^{(ij)}\lambda^{n-2} + \cdots + c_r^{(ij)}\lambda^{n-r}$ for all $i = 1, \ldots, n$ and $j = 1, \ldots, m$. By substituting these values in the matrix from (35), we obtain

$$\mathbf{A}^{\dagger}_{M,N} = \lim_{\lambda \to 0}\begin{pmatrix} \frac{c_1^{(11)}\lambda^{n-1}+\ldots+c_r^{(11)}\lambda^{n-r}}{\alpha^n+d_1\lambda^{n-1}+\ldots+d_r\lambda^{n-r}} & \ldots & \frac{c_1^{(1m)}\lambda^{n-1}+\ldots+c_r^{(1m)}\lambda^{n-r}}{\lambda^n+d_1\lambda^{n-1}+\ldots+d_r\lambda^{n-r}} \\ \ldots & \ldots & \ldots \\ \frac{c_1^{(n1)}\lambda^{n-1}+\ldots+c_r^{(n1)}\lambda^{n-r}}{\lambda^n+d_1\lambda^{n-1}+\ldots+d_r\lambda^{n-r}} & \ldots & \frac{c_1^{(nm)}\lambda^{n-1}+\ldots+c_r^{(nm)}\lambda^{n-r}}{\lambda^n+d_1\lambda^{n-1}+\ldots+d_r\lambda^{n-r}} \end{pmatrix} = \begin{pmatrix} \frac{c_r^{(11)}}{d_r} & \ldots & \frac{c_r^{(1m)}}{d_r} \\ \ldots & \ldots & \ldots \\ \frac{c_r^{(n1)}}{d_r} & \ldots & \frac{c_r^{(nm)}}{d_r} \end{pmatrix}.$$

Here

$$c_r^{(ij)} = \sum_{\beta \in J_{r,\,n}\{i\}} \operatorname{cdet}_i \left(\left(\mathbf{A}^\sharp \mathbf{A} \right)_{.\,i} \left(\mathbf{a}^\sharp_{.j} \right) \right)^{\beta}_{\beta}$$

and $d_r = \sum_{\beta \in J_{r,\,n}} \left| \left(\mathbf{A}^\sharp \mathbf{A} \right)^{\beta}_{\beta} \right|$. Thus, we have obtained the determinantal representation of $\mathbf{A}^\dagger_{M,N}$ by (33).

The determinantal representation of $\mathbf{A}^\dagger_{M,N}$ by (34) can be proved similarly. □

Corollary 4.1. *If* $\operatorname{rank}\mathbf{A} = n < m$, *and* $(\mathbf{A}^\sharp \mathbf{A})$ *is Hermitian, then we get the following representation of* $\mathbf{A}^\dagger_{M,N} = \left(a^\ddagger_{ij} \right) \in \mathbb{H}^{n \times m}$,

$$a^\ddagger_{ij} = \frac{\operatorname{cdet}_i (\mathbf{A}^\sharp \mathbf{A})_{.\,i} \left(\mathbf{a}^\sharp_{.\,j} \right)}{\det (\mathbf{A}^\sharp \mathbf{A})}, \tag{36}$$

or the determinantal representation (33) can be applicable as well.

Proof. By Corollary 3.1, $\mathbf{A}^\dagger_{M,N} = \left(\mathbf{A}^\sharp \mathbf{A} \right)^{-1} \mathbf{A}^\sharp$. Considering Hermitian $\left(\mathbf{A}^\sharp \mathbf{A} \right)^{-1}$ as a left inverse by (7), we obtain

$$a^\ddagger_{ij} = \frac{\sum_k L_{ki} a^\sharp_{kj}}{\det (\mathbf{A}^\sharp \mathbf{A})},$$

where L_{ki} is the (ki)th cofactor of $\det(\mathbf{A}^\sharp \mathbf{A}) = \operatorname{cdet}_i(\mathbf{A}^\sharp \mathbf{A})$. By its definition, $\sum_k L_{ki} a^\sharp_{kj} = \operatorname{cdet}_i (\mathbf{A}^\sharp \mathbf{A})_{.i} \left(\mathbf{a}^\sharp_{.j} \right)$. From this, (36) follows immediately. □

Corollary 4.2. *If* $\operatorname{rank}\mathbf{A} = m$, *and* $(\mathbf{A}\mathbf{A}^\sharp)$ *is Hermitian, then*

$$a^\ddagger_{ij} = \frac{\operatorname{rdet}_j (\mathbf{A}\mathbf{A}^\sharp)_{j.} \left(\mathbf{a}^\sharp_{i.} \right)}{\det (\mathbf{A}\mathbf{A}^\sharp)}. \tag{37}$$

or the determinantal representation (34) can be applicable as well.

Proof. By Corollary 3.1, $\mathbf{A}^\dagger_{M,N} = \mathbf{A}^\sharp \left(\mathbf{A}\mathbf{A}^\sharp \right)^{-1}$. Considering $\left(\mathbf{A}\mathbf{A}^\sharp \right)^{-1}$ as a right inverse by (6), we get the following representation

$$a^\ddagger_{ij} = \frac{\sum_k a^\sharp_{ik} R_{jk}}{\det (\mathbf{A}^\sharp \mathbf{A})},$$

where R_{jk} is the (jk)th cofactor of $\det(\mathbf{A}\mathbf{A}^{\sharp}) = \operatorname{rdet}_j(\mathbf{A}\mathbf{A}^{\sharp})$. By its definition, $\sum_k a^{\sharp}_{ik} R_{jk} = \operatorname{rdet}_j(\mathbf{A}\mathbf{A}^{\sharp})_{j.}(\mathbf{a}^{\sharp}_{i.})$. From this, (37) follows immediately. □

We also can obtain determinantal representations of the projection matrices $\mathbf{A}^{\dagger}_{M,N}\mathbf{A}$ and $\mathbf{A}\mathbf{A}^{\dagger}_{M,N}$ in the following corollaries.

Corollary 4.3. *If* $\mathbf{A} \in \mathbb{H}^{m\times n}_r$ *and* $\mathbf{A}^{\sharp}\mathbf{A}$ *is Hermitian, then the projection matrix* $\mathbf{A}^{\dagger}_{M,N}\mathbf{A} =: \mathbf{P} = (p_{ij})_{n\times n}$ *possess the following determinantal representation,*

(i) if $r < \min\{m,n\}$ *or* $r = m < n$,

$$p_{ij} = \frac{\sum\limits_{\beta\in J_{r,\,n}\{i\}} \operatorname{cdet}_i\left(\left(\mathbf{A}^{\sharp}\mathbf{A}\right)_{.\,i}(\mathbf{d}_{.j})\right)^{\beta}_{\beta}}{\sum\limits_{\beta\in J_{r,\,n}} \left|\left(\mathbf{A}^{\sharp}\mathbf{A}\right)^{\beta}_{\beta}\right|}, \tag{38}$$

where $\mathbf{d}_{.j}$ *is the* j*th column of* $\mathbf{A}^{\sharp}\mathbf{A} \in \mathbb{H}^{n\times n}$ *and for all* $i,j = 1,\dots,n$;

(ii) if $r = n$,

$$p_{ij} = \frac{\operatorname{cdet}_i(\mathbf{A}^{\sharp}\mathbf{A})_{.\,i}(\mathbf{d}_{.j})}{\det(\mathbf{A}^{\sharp}\mathbf{A})}.$$

Proof.

(i) If $\mathbf{A} \in \mathbb{H}^{m\times n}_r$, $\mathbf{A}^{\sharp}\mathbf{A}$ is Hermitian, and $r < \min\{m,n\}$ or $r = m < n$, we can represent $\mathbf{A}^{\dagger}$ by (33). Right-multiplying it by $\mathbf{A}$ gives the following presentation of an entry p_{ij} of $\mathbf{A}^{\dagger}_{M,N}\mathbf{A} =: \mathbf{P} = (p_{ij})_{n\times n}$,

$$p_{ij} = \sum_k \frac{\sum\limits_{\beta\in J_{r,\,n}\{i\}} \operatorname{cdet}_i\left(\left(\mathbf{A}^{\sharp}\mathbf{A}\right)_{.\,i}\left(\mathbf{a}^{\sharp}_{.k}\right)\right)^{\beta}_{\beta}}{\sum\limits_{\beta\in J_{r,\,n}} \left|(\mathbf{A}^{\sharp}\mathbf{A})^{\beta}_{\beta}\right|} \cdot a_{kj} =$$

$$= \frac{\sum\limits_{\beta\in J_{r,\,n}\{i\}} \sum\limits_k \operatorname{cdet}_i\left(\left(\mathbf{A}^{\sharp}\mathbf{A}\right)_{.\,i}\left(\mathbf{a}^{\sharp}_{.k}\right)\right)^{\beta}_{\beta} \cdot a_{kj}}{\sum\limits_{\beta\in J_{r,\,n}} \left|(\mathbf{A}^{\sharp}\mathbf{A})^{\beta}_{\beta}\right|}.$$

Since $\sum\limits_k \mathbf{a}^{\sharp}_{.k} a_{kj} = \mathbf{d}_{.j}$, where $\mathbf{d}_{.j}$ denote the jth column of $\mathbf{A}^{\sharp}\mathbf{A} \in \mathbb{H}^{n\times n}$, then it follows (38).

(ii) If $\mathbf{A} \in \mathbb{H}_r^{m\times n}$, $\mathbf{A}^\sharp\mathbf{A}$ is Hermitian, and $r = n$, we can represent $\mathbf{A}^\dagger$ by (36). Right-multiplying it by $\mathbf{A}$ gives the following presentation of an entry p_{ij} of $\mathbf{A}^\dagger_{M,N}\mathbf{A} =: \mathbf{P} = (p_{ij})_{n\times n}$,

$$p_{ij} = \sum_{k=1}^{n} \frac{\mathrm{cdet}_i(\mathbf{A}^\sharp\mathbf{A})_{.\,i}\left(\mathbf{a}^\sharp_{.j}\right)}{\det(\mathbf{A}^\sharp\mathbf{A})} \cdot a_{kj} = \frac{\mathrm{cdet}_i(\mathbf{A}^\sharp\mathbf{A})_{.\,i}\,(\mathbf{d}_{.j})}{\det(\mathbf{A}^\sharp\mathbf{A})}.$$

□

The following corollary can be proved by analogy.

Corollary 4.4. *If $\mathbf{A} \in \mathbb{H}_r^{m\times n}$ and $(\mathbf{A}\mathbf{A}^\sharp) \in \mathbb{H}^{m\times m}$ is Hermitian, then the projection matrix $\mathbf{A}\mathbf{A}^\dagger_{M,N} =: \mathbf{Q} = (q_{ij})_{m\times m}$ possess the following determinantal representation,*

(i) if $r < \min\{m,n\}$ or $r = n < m$,

$$q_{ij} = \frac{\sum\limits_{\alpha\in I_{r,\,m}\{j\}} \mathrm{rdet}_j\left((\mathbf{A}\,\mathbf{A}^\sharp)_{j.}\,(\mathbf{g}_{i.})\right)\,{}^{\alpha}_{\alpha}}{\sum\limits_{\alpha\in I_{r,\,m}} |(\mathbf{A}\mathbf{A}^\sharp)\,{}^{\alpha}_{\alpha}|},$$

where $\mathbf{g}_{i.}$ is the ith row of $(\mathbf{A}\mathbf{A}^\sharp) \in \mathbb{H}^{m\times m}$ and for all $i,j = 1,\ldots,m$.

(ii) if $r = m$,

$$q_{ij} = \frac{\mathrm{rdet}_j(\mathbf{A}\mathbf{A}^\sharp)_{j.}\,(\mathbf{g}_{i.})}{\det(\mathbf{A}\mathbf{A}^\sharp)}.$$

4.2. The Case of Non-Hermitian $\mathbf{A}^\sharp\mathbf{A}$ and $\mathbf{A}\mathbf{A}^\sharp$

In this subsection we derive determinantal representations of the weighted Moore-Penrose inverse of $\mathbf{A} \in \mathbb{H}^{m\times n}$ when $(\mathbf{A}\mathbf{A}^\sharp) \in \mathbb{H}^{m\times m}$ and $(\mathbf{A}^\sharp\mathbf{A}) \in \mathbb{H}^{n\times n}$ are non-Hermitian.

First, let $(\mathbf{A}^\sharp\mathbf{A}) \in \mathbb{H}^{n\times n}$ be non-Hermitian and $\mathrm{rank}(\mathbf{A}^\sharp\mathbf{A}) < n$. Consider $(\lambda\mathbf{I} + \mathbf{A}^\sharp\mathbf{A})^{-1}\mathbf{A}^\sharp$. We have,

$$\begin{gathered}(\lambda\mathbf{I}+\mathbf{A}^\sharp\mathbf{A})^{-1}\mathbf{A}^\sharp = (\lambda\mathbf{I}+\mathbf{N}^{-1}\mathbf{A}^*\mathbf{M}\mathbf{A})^{-1}\mathbf{A}^\sharp = (\mathbf{N}^{-1}(\lambda\mathbf{N}+\mathbf{A}^*\mathbf{M}\mathbf{A}))^{-1}\mathbf{A}^\sharp = \\ (\lambda\mathbf{N}+\mathbf{A}^*\mathbf{M}\mathbf{A})^{-1}\mathbf{A}^*\mathbf{M} = \mathbf{N}^{-\frac{1}{2}}(\lambda+\mathbf{N}^{-\frac{1}{2}}\mathbf{A}^*\mathbf{M}\mathbf{A}\mathbf{N}^{-\frac{1}{2}})^{-1}\mathbf{N}^{-\frac{1}{2}}\mathbf{A}^*\mathbf{M} = \\ \mathbf{N}^{-\frac{1}{2}}\left(\lambda+\left(\mathbf{M}^{\frac{1}{2}}\mathbf{A}\mathbf{N}^{-\frac{1}{2}}\right)^*\mathbf{M}^{\frac{1}{2}}\mathbf{A}\mathbf{N}^{-\frac{1}{2}}\right)^{-1}\left(\mathbf{N}^{-\frac{1}{2}}\mathbf{A}^*\mathbf{M}^{\frac{1}{2}}\right)\mathbf{M}^{\frac{1}{2}} \qquad (39)\end{gathered}$$

Since by Lemma 2.2

$$\lim_{\lambda\to 0}\left(\lambda+\left(\mathbf{M}^{\frac{1}{2}}\mathbf{A}\mathbf{N}^{-\frac{1}{2}}\right)^{*}\mathbf{M}^{\frac{1}{2}}\mathbf{A}\mathbf{N}^{-\frac{1}{2}}\right)^{-1}\left(\mathbf{N}^{-\frac{1}{2}}\mathbf{A}^{*}\mathbf{M}^{\frac{1}{2}}\right)=\left(\mathbf{M}^{\frac{1}{2}}\mathbf{A}\mathbf{N}^{-\frac{1}{2}}\right)^{\dagger}, \tag{40}$$

then combining (40) and (39), we obtain the well-known representation of the weighted Moore-Penrose inverse(see, e.g., [40]),

$$\mathbf{A}^{\dagger}_{M,N}=\mathbf{N}^{-\frac{1}{2}}\left(\mathbf{M}^{\frac{1}{2}}\mathbf{A}\mathbf{N}^{-\frac{1}{2}}\right)^{\dagger}\mathbf{M}^{\frac{1}{2}}. \tag{41}$$

Denote $\mathbf{M}^{\frac{1}{2}}=\left(m_{ij}^{(\frac{1}{2})}\right)$, $\mathbf{N}^{-\frac{1}{2}}=\left(n_{ij}^{(-\frac{1}{2})}\right)$, and $\widetilde{\mathbf{A}}:=\mathbf{M}^{\frac{1}{2}}\mathbf{A}\mathbf{N}^{-\frac{1}{2}}=(\widetilde{a}_{ij})\in\mathbb{H}^{m\times n}$, then $\mathbf{N}^{-\frac{1}{2}}\mathbf{A}^{*}\mathbf{M}^{\frac{1}{2}}=\widetilde{\mathbf{A}}^{*}=\left(\widetilde{a}^{*}_{ij}\right)$, $\left(\mathbf{M}^{\frac{1}{2}}\mathbf{A}\mathbf{N}^{-\frac{1}{2}}\right)^{\dagger}=\widetilde{\mathbf{A}}^{\dagger}=\left(\widetilde{a}^{\dagger}_{ij}\right)$. By determinantal representing (13) for $\widetilde{\mathbf{A}}^{\dagger}$, we obtain

$$\widetilde{a}^{\dagger}_{ij}=\frac{\sum\limits_{\beta\in J_{r,\,n}\{i\}}\operatorname{cdet}_i\left(\left(\left(\mathbf{M}^{\frac{1}{2}}\mathbf{A}\mathbf{N}^{-\frac{1}{2}}\right)^{*}\left(\mathbf{M}^{\frac{1}{2}}\mathbf{A}\mathbf{N}^{-\frac{1}{2}}\right)\right)_{.\,i}\left(\mathbf{m}^{\frac{1}{2}}\mathbf{a}\mathbf{n}^{-\frac{1}{2}}\right)^{*}_{.j}\right)^{\beta}_{\beta}}{\sum\limits_{\beta\in J_{r,\,n}}\left|\left(\left(\mathbf{M}^{\frac{1}{2}}\mathbf{A}\mathbf{N}^{-\frac{1}{2}}\right)^{*}\left(\mathbf{M}^{\frac{1}{2}}\mathbf{A}\mathbf{N}^{-\frac{1}{2}}\right)\right)^{\beta}_{\beta}\right|}=$$

$$\frac{\sum\limits_{\beta\in J_{r,\,n}\{i\}}\operatorname{cdet}_i\left(\left(\left(\mathbf{M}^{\frac{1}{2}}\mathbf{A}\mathbf{N}^{-\frac{1}{2}}\right)^{*}\left(\mathbf{M}^{\frac{1}{2}}\mathbf{A}\mathbf{N}^{-\frac{1}{2}}\right)\right)_{.\,i}\left(\mathbf{n}^{-\frac{1}{2}}\mathbf{a}^{*}\mathbf{m}^{\frac{1}{2}}\right)_{.j}\right)^{\beta}_{\beta}}{\sum\limits_{\beta\in J_{r,\,n}}\left|\left(\left(\mathbf{M}^{\frac{1}{2}}\mathbf{A}\mathbf{N}^{-\frac{1}{2}}\right)^{*}\left(\mathbf{M}^{\frac{1}{2}}\mathbf{A}\mathbf{N}^{-\frac{1}{2}}\right)\right)^{\beta}_{\beta}\right|}$$

where $\left(\mathbf{m}^{\frac{1}{2}}\mathbf{a}\mathbf{n}^{-\frac{1}{2}}\right)^{*}_{.j}$ denote the j-th column of $\left(\mathbf{M}^{\frac{1}{2}}\mathbf{A}\mathbf{N}^{-\frac{1}{2}}\right)^{*}$, $\left(\mathbf{n}^{-\frac{1}{2}}\mathbf{a}^{*}\mathbf{m}^{\frac{1}{2}}\right)_{.j}$ denote the j-th column of $\left(\mathbf{N}^{-\frac{1}{2}}\mathbf{A}^{*}\mathbf{M}^{\frac{1}{2}}\right)$ for all $j=1,\ldots,m$. Since $\sum_l\left(\mathbf{n}^{-\frac{1}{2}}\mathbf{a}^{*}\mathbf{m}^{\frac{1}{2}}\right)_{.\,l}m_{lj}^{\frac{1}{2}}=\left(\mathbf{n}^{-\frac{1}{2}}\mathbf{a}^{*}\mathbf{m}\right)_{.j}$, then for the weighted Moore-

Penrose inverse $\mathbf{A}^{\dagger}_{M,N} = \left(a^{\ddagger}_{ij}\right) \in \mathbb{H}^{n\times m}$, we have

$$a^{\ddagger}_{ij} = \sum_{k}^{n}\sum_{l}^{m} n_{ik}^{-\frac{1}{2}}\widetilde{a}^{\dagger}_{kl}m_{lj}^{\frac{1}{2}} =$$

$$\frac{\sum\limits_{k} n_{ik}^{-\frac{1}{2}}\cdot \sum\limits_{\beta\in J_{r,\,n}\{i\}} \mathrm{cdet}_k\left(\left(\left(\mathbf{M}^{\frac{1}{2}}\mathbf{A}\mathbf{N}^{-\frac{1}{2}}\right)^{*}\left(\mathbf{M}^{\frac{1}{2}}\mathbf{A}\mathbf{N}^{-\frac{1}{2}}\right)\right)_{.\,k}\left(\mathbf{n}^{-\frac{1}{2}}\mathbf{a}^{*}\mathbf{m}\right)_{.j}\right)^{\beta}_{\beta}}{\sum\limits_{\beta\in J_{r,\,n}}\left|\left(\left(\mathbf{M}^{\frac{1}{2}}\mathbf{A}\mathbf{N}^{-\frac{1}{2}}\right)^{*}\left(\mathbf{M}^{\frac{1}{2}}\mathbf{A}\mathbf{N}^{-\frac{1}{2}}\right)\right)^{\beta}_{\beta}\right|} =$$

$$\frac{\sum\limits_{k} n_{ik}^{(-\frac{1}{2})}\sum\limits_{\beta\in J_{r,\,n}\{i\}} \mathrm{cdet}_k\left(\left(\widetilde{\mathbf{A}}^{*}\widetilde{\mathbf{A}}\right)_{.\,k}\left(\widehat{a}_{ij}\right)\right)^{\beta}_{\beta}}{\sum\limits_{\beta\in J_{r,\,n}}\left|\left(\widetilde{\mathbf{A}}^{*}\widetilde{\mathbf{A}}\right)^{\beta}_{\beta}\right|},$$

where $\widehat{\mathbf{a}}_{.j} := \left(\mathbf{n}^{-\frac{1}{2}}\mathbf{a}^{*}\mathbf{m}\right)_{.j}$ denote jth column of $\widehat{\mathbf{A}} = (\widehat{a}_{ij}) := \left(\mathbf{N}^{-\frac{1}{2}}\mathbf{A}^{*}\mathbf{M}\right)$ for all $i = 1,\ldots,n$, $j = 1,\ldots,m$.

If $\mathrm{rank}(\mathbf{A}^{\sharp}\mathbf{A}) = n$, then by Corollary 3.1, $\mathbf{A}^{\dagger}_{M,N} = (\mathbf{A}^{\sharp}\mathbf{A})^{-1}\mathbf{A}^{\sharp} =: \left(a^{\ddagger}_{ij}\right)$. So, $\mathbf{A}^{\dagger}_{M,N} = \left(\mathbf{N}^{-1}\mathbf{A}^{*}\mathbf{M}\mathbf{A}\right)^{-1}\mathbf{N}^{-1}\mathbf{A}^{*}\mathbf{M} = \left(\mathbf{A}^{*}\mathbf{M}\mathbf{A}\right)^{-1}\mathbf{A}^{*}\mathbf{M}$. Since $\mathbf{A}^{*}\mathbf{M}\mathbf{A}$ is Hermitian, then we can use the determinantal representation of a Hermitian inverse matrix (7). Denote $\mathbf{A}^{*}\mathbf{M} =: \hat{\mathbf{A}} = (\hat{a})_{ij} \in \mathbb{H}^{n\times m}$. So, we have

$$a^{\ddagger}_{ij} = \frac{\sum_{k=1}^{n} L_{ki}\hat{a}_{kj}}{\det(\mathbf{A}^{*}\mathbf{M}\mathbf{A})} = \frac{\mathrm{cdet}_i(\mathbf{A}^{*}\mathbf{M}\mathbf{A})_{.i}\left(\hat{\mathbf{a}}_{.j}\right)}{\det(\mathbf{A}^{*}\mathbf{M}\mathbf{A})}.$$

where $\hat{\mathbf{a}}_{.j}$ is the jth column of $\mathbf{A}^{*}\mathbf{M}$ for all $j = 1,\ldots,m$.

Now, let $(\mathbf{A}\mathbf{A}^{\sharp}) \in \mathbb{H}^{m\times m}$ be non-Hermitian and $\mathrm{rank}(\mathbf{A}\mathbf{A}^{\sharp}) < m$. By determinantal representing (14) for $\widetilde{\mathbf{A}}^{\dagger}$, we similarly obtain

$$\widetilde{a}^{\dagger}_{ij} = \frac{\sum\limits_{\alpha\in I_{r,\,m}\{j\}} \mathrm{rdet}_j\left(\left(\left(\mathbf{M}^{\frac{1}{2}}\mathbf{A}\mathbf{N}^{-\frac{1}{2}}\right)\left(\mathbf{M}^{\frac{1}{2}}\mathbf{A}\mathbf{N}^{-\frac{1}{2}}\right)\right)^{*}_{j.}\left(\mathbf{n}^{-\frac{1}{2}}\mathbf{a}^{*}\mathbf{m}^{\frac{1}{2}}\right)_{i.}\right)^{\alpha}_{\alpha}}{\sum\limits_{\alpha\in I_{r,\,m}}\left|\left(\left(\mathbf{M}^{\frac{1}{2}}\mathbf{A}\mathbf{N}^{-\frac{1}{2}}\right)\left(\mathbf{M}^{\frac{1}{2}}\mathbf{A}\mathbf{N}^{-\frac{1}{2}}\right)^{*}\right)^{\alpha}_{\alpha}\right|},$$

where $\left(\mathbf{n}^{-\frac{1}{2}}\mathbf{a}^{*}\mathbf{m}^{\frac{1}{2}}\right)_{i.}$ denote the ith row of $\left(\mathbf{N}^{-\frac{1}{2}}\mathbf{A}^{*}\mathbf{M}^{\frac{1}{2}}\right)$ for all $i = \overline{1,n}$.

Since $\sum\limits_k n_{ik}^{-\frac{1}{2}}\left(\mathbf{n}^{-\frac{1}{2}}\mathbf{a}^*\mathbf{m}^{\frac{1}{2}}\right)_{k.} = \left(\mathbf{n}^{-\frac{1}{2}}\mathbf{a}^*\mathbf{m}^{\frac{1}{2}}\right)_{i.}$, then we get

$$a_{ij}^{\ddagger} = \sum_k^n \sum_l^m n_{ik}^{-\frac{1}{2}} \widetilde{a}_{kl}^{\dagger} m_{lj}^{\frac{1}{2}} =$$

$$\sum_l \frac{\sum\limits_{\alpha\in I_{r,\,m}\{l\}} \operatorname{rdet}_l\left(\left(\left(\mathbf{M}^{\frac{1}{2}}\mathbf{A}\mathbf{N}^{-\frac{1}{2}}\right)\left(\mathbf{M}^{\frac{1}{2}}\mathbf{A}\mathbf{N}^{-\frac{1}{2}}\right)^*\right)_{l.}\left(\mathbf{n}^{-1}\mathbf{a}^*\mathbf{m}^{\frac{1}{2}}\right)_{i.}\right)_{\alpha}^{\alpha}}{\sum\limits_{\alpha\in I_{r,\,m}}\left|\left(\left(\mathbf{M}^{\frac{1}{2}}\mathbf{A}\mathbf{N}^{-\frac{1}{2}}\right)\left(\mathbf{M}^{\frac{1}{2}}\mathbf{A}\mathbf{N}^{-\frac{1}{2}}\right)^*\right)_{\alpha}^{\alpha}\right|} \cdot m_{lj}^{\frac{1}{2}} =$$

$$\frac{\sum\limits_l \sum\limits_{\alpha\in I_{r,\,m}\{l\}} \operatorname{rdet}_l\left(\left(\widetilde{\mathbf{A}}\widetilde{\mathbf{A}}^*\right)_{l.}(\widehat{\mathbf{a}}_{i.})\right)_{\alpha}^{\alpha} \cdot m_{lj}^{\frac{1}{2}}}{\sum\limits_{\alpha\in I_{r,\,m}}\left|\left(\widetilde{\mathbf{A}}\widetilde{\mathbf{A}}^*\right)_{\alpha}^{\alpha}\right|},$$

where, in this case, $\widehat{\mathbf{a}}_{i.} := \left(\mathbf{n}^{-1}\mathbf{a}^*\mathbf{m}^{\frac{1}{2}}\right)_{i.}$ denote ith row of $\widehat{\mathbf{A}} = (\widehat{a}_{ij}) := \left(\mathbf{N}^{-1}\mathbf{A}^*\mathbf{M}^{\frac{1}{2}}\right)$ for all $i = 1,\ldots,n$, $j = 1,\ldots,m$.

If $\operatorname{rank}(\mathbf{A}\mathbf{A}^{\sharp}) = m$, then by Corollary 3.1, $\mathbf{A}_{M,N}^{\dagger} = \mathbf{A}^{\sharp}(\mathbf{A}\mathbf{A}^{\sharp})^{-1}$. So, $\mathbf{A}_{M,N}^{\dagger} = \mathbf{N}^{-1}\mathbf{A}^*\mathbf{M}\left(\mathbf{A}\mathbf{N}^{-1}\mathbf{A}^*\mathbf{M}\right)^{-1} = \mathbf{N}^{-1}\mathbf{A}^*\left(\mathbf{A}\mathbf{N}^{-1}\mathbf{A}^*\right)^{-1}$. Since $\mathbf{A}\mathbf{N}^{-1}\mathbf{A}^*$ is Hermitian and full-rank, then we can use the determinantal representation of a Hermitian inverse matrix (6). Denote $\mathbf{N}^{-1}\mathbf{A}^* =: (\widehat{a})_{ij} \in \mathbb{H}^{n\times m}$. So, we have

$$a_{ij}^{\ddagger} = \frac{\sum_{k=1}^n \breve{a}_{ik}R_{jk}}{\det(\mathbf{A}\mathbf{N}^{-1}\mathbf{A}^*)} = \frac{\operatorname{rdet}_j(\mathbf{A}\mathbf{N}^{-1}\mathbf{A}^*)_{j.}(\widehat{\mathbf{a}}_{i.})}{\det(\mathbf{A}\mathbf{N}^{-1}\mathbf{A}^*)},$$

where $\widehat{\mathbf{a}}_{i.}$ is the ith row of $\mathbf{N}^{-1}\mathbf{A}^*$ for all $i = 1,\ldots,n$.

Thus, we have proved the following theorem.

Theorem 4.2. *Let* $\mathbf{A} \in \mathbb{H}_r^{m\times n}$.

(i) If $\mathbf{A}^{\sharp}\mathbf{A}$ *is non-Hermitian, then the weighted Moore-Penrose inverse* $\mathbf{A}_{M,N}^{\dagger} = \left(a_{ij}^{\ddagger}\right) \in \mathbb{H}^{n\times m}$ *possess the determinantal representations*

(a) if $r < n$

$$a_{ij}^{\ddagger} = \frac{\sum\limits_k n_{ik}^{(-\frac{1}{2})} \sum\limits_{\beta\in J_{r,\,n}\{i\}} \operatorname{cdet}_k\left(\left(\widetilde{\mathbf{A}}^*\widetilde{\mathbf{A}}\right)_{.k}(\widehat{\mathbf{a}}_{.j})\right)_{\beta}^{\beta}}{\sum\limits_{\beta\in J_{r,\,n}}\left|\left(\widetilde{\mathbf{A}}^*\widetilde{\mathbf{A}}\right)_{\beta}^{\beta}\right|}, \tag{42}$$

where $\widehat{\mathbf{a}}_{.j}$ *is the* j*th column of* $\mathbf{N}^{-\frac{1}{2}}\mathbf{A}^*\mathbf{M}$*;*

(b) if $r = n$

$$a_{ij}^{\ddagger} = \frac{\operatorname{cdet}_i(\mathbf{A}^*\mathbf{M}\mathbf{A})_{.i}(\widehat{\mathbf{a}}_{.j})}{\det(\mathbf{A}^*\mathbf{M}\mathbf{A})}, \tag{43}$$

where $\widehat{\mathbf{a}}_{.j}$ *is the* j*th column of* $\mathbf{A}^*\mathbf{M}$ *for all* $j = 1, \ldots, m$*.*

(ii) If $\mathbf{A}\mathbf{A}^{\sharp}$ *is non-Hermitian, then* $\mathbf{A}_{M,N}^{\dagger} = \left(a_{ij}^{\ddagger}\right)$ *possess the determinantal representation*

(a) if $r < m$*,*

$$a_{ij}^{\ddagger} = \frac{\sum\limits_{l}\sum\limits_{\alpha\in I_{r,\,m}\{l\}} \operatorname{rdet}_l\left(\left(\widetilde{\mathbf{A}}\widetilde{\mathbf{A}}^*\right)_{l.}(\widehat{\mathbf{a}}_{i.})\right)_{\alpha}^{\alpha} \cdot m_{lj}^{(\frac{1}{2})}}{\sum\limits_{\alpha\in I_{r,\,m}} \left|\left(\widetilde{\mathbf{A}}\widetilde{\mathbf{A}}^*\right)_{\alpha}^{\alpha}\right|}, \tag{44}$$

where $\widehat{\mathbf{a}}_{i.}$ *is the* i*th row of* $\mathbf{N}^{-1}\mathbf{A}^*\mathbf{M}^{\frac{1}{2}}$*;*

(b) if $r = m$*,*

$$a_{ij}^{\ddagger} = \frac{\operatorname{rdet}_j(\mathbf{A}\mathbf{N}^{-1}\mathbf{A}^*)_{j.}(\widehat{\mathbf{a}}_{i.})}{\det(\mathbf{A}\mathbf{N}^{-1}\mathbf{A}^*)}. \tag{45}$$

where $\widehat{\mathbf{a}}_{i.}$ *is the* i*th row of* $\mathbf{N}^{-1}\mathbf{A}^*$ *for all* $i = 1, \ldots, n$*.*

Remark 4.1. *To give determinantal representations of* $\mathbf{A}_{M,N}^{\dagger}$ *over the complex field it is enough in Theorem 4.1 substitute all row-column determinants by usual determinants.*

5. Cramer's Rule for Systems of Quaternion Linear Equations with Restrictions

Consider a right system of linear equations over the quaternion skew field,

$$\mathbf{A}\mathbf{x} = \mathbf{b} \tag{46}$$

where $\mathbf{A} \in \mathbb{H}^{m\times n}$ is a coefficient matrix, $\mathbf{b} \in \mathbb{H}^{m\times 1}$ is a column of constants, and $\mathbf{x} \in \mathbb{H}^{n\times 1}$ is a unknown column. Due to [16], we have the following theorem that characterizes the weighted Moore-Penrose solution of (46).

Theorem 5.1. *[16] The right system of linear equations (46) with restriction*

$$\mathbf{x} \in \mathcal{R}_r(\mathbf{A}^\sharp) \tag{47}$$

has the unique solution $\tilde{\mathbf{x}} = \mathbf{A}^\dagger_{M,N}\mathbf{b}$.

The following theorems give analogs of Cramer's rule for solutions of the system (46) with the restriction (47).

Theorem 5.2. *Let* $\mathbf{A} \in \mathbb{H}^{m\times n}$ *and* $\mathbf{A}^\sharp\mathbf{A} \in \mathbb{H}^{n\times n}$ *be Hermitian.*

(i) If $\operatorname{rank}\mathbf{A} = r \le m < n$*, then the solution* $\tilde{\mathbf{x}} = (\tilde{x}_1, \ldots, \tilde{x}_n)^T \in \mathbb{H}^{n\times 1}$ *of (46) possess the following determinantal representations*

$$\tilde{x}_i = \frac{\sum\limits_{\beta\in J_{r,\,n}\{i\}} \operatorname{cdet}_i\left(\left(\mathbf{A}^\sharp\mathbf{A}\right)_{.\,i}(\mathbf{f})\right)^\beta_\beta}{\sum\limits_{\beta\in J_{r,\,n}} \left|\left(\mathbf{A}^\sharp\mathbf{A}\right)^\beta_\beta\right|}, \tag{48}$$

where $\mathbf{f} = \mathbf{A}^\sharp\,\mathbf{b} \in \mathbb{H}^{n\times 1}$.

(ii) If $\operatorname{rank}\mathbf{A} = n$*, then*

$$\tilde{x}_i = \frac{\operatorname{cdet}_i\left(\mathbf{A}^\sharp\mathbf{A}\right)_{.i}(\mathbf{f})}{\det\mathbf{A}^\sharp\mathbf{A}}. \tag{49}$$

Proof. (i) If $\operatorname{rank}\mathbf{A} = r \le m < n$, then by Theorem 4.1 we can represent $\mathbf{A}^\dagger_{M,N}$ by (33). By component-wise of $\tilde{\mathbf{x}} = \mathbf{A}^\dagger_{M,N}\mathbf{b}$, we have

$$\tilde{x}_i = \sum_{j=1}^{m} \frac{\sum\limits_{\beta\in J_{r,\,n}\{i\}} \operatorname{cdet}_i\left(\left(\mathbf{A}^\sharp\mathbf{A}\right)_{.\,i}\left(\mathbf{a}^\sharp_{.j}\right)\right)^\beta_\beta}{\sum\limits_{\beta\in J_{r,\,n}} \left|\left(\mathbf{A}^\sharp\mathbf{A}\right)^\beta_\beta\right|} \cdot b_j =$$

$$= \frac{\sum\limits_{\beta\in J_{r,\,n}\{i\}} \sum\limits_{j} \operatorname{cdet}_i\left(\left(\mathbf{A}^\sharp\mathbf{A}\right)_{.\,i}\left(\mathbf{a}^\sharp_{.j}\right)\right)^\beta_\beta \cdot b_j}{\sum\limits_{\beta\in J_{r,\,n}} \left|\left(\mathbf{A}^\sharp\mathbf{A}\right)^\beta_\beta\right|} = \frac{\sum\limits_{\beta\in J_{r,\,n}\{i\}} \operatorname{cdet}_i\left(\left(\mathbf{A}^\sharp\mathbf{A}\right)_{.\,i}(\mathbf{f})\right)^\beta_\beta}{\sum\limits_{\beta\in J_{r,\,n}} \left|\left(\mathbf{A}^\sharp\mathbf{A}\right)^\beta_\beta\right|},$$

where $\mathbf{f} = \mathbf{A}^\sharp\mathbf{b}$ and for all $i = 1, \ldots, n$.
(ii) If $\operatorname{rank}\mathbf{A} = n$, then $\mathbf{A}^\dagger_{M,N}$ can be represented by (36). Representing $\mathbf{A}^\dagger\mathbf{b}$ by component-wise directly gives (49). □

Theorem 5.3. *Let $\mathbf{A} \in \mathbb{H}^{m\times n}$ and $\mathbf{A}^\sharp\mathbf{A} \in \mathbb{H}^{n\times n}$ be non-Hermitian.*

(i) If $\operatorname{rank}\mathbf{A} = r \le m < n$*, then the solution* $\tilde{\mathbf{x}} = (\tilde{x}_1, \ldots, \tilde{x}_n)^T \in \mathbb{H}^{n\times 1}$ *of (46) possess the following determinantal representation*

$$\tilde{x}_i = \frac{\sum_k n_{ik}^{(-\frac{1}{2})} \sum_{\beta\in J_{r,\,n}\{i\}} \operatorname{cdet}_k\left(\left(\widetilde{\mathbf{A}}^*\widetilde{\mathbf{A}}\right)_{.k}(\mathbf{f})\right)_\beta^\beta}{\sum_{\beta\in J_{r,\,n}} \left|\left(\widetilde{\mathbf{A}}^*\widetilde{\mathbf{A}}\right)_\beta^\beta\right|},$$

where $\mathbf{f} = \left(\mathbf{N}^{-\frac{1}{2}}\mathbf{A}^\sharp\mathbf{M}\right)\mathbf{b} \in \mathbb{H}^{n\times 1}$ *and* $n_{ik}^{(-\frac{1}{2})}$ *is the* (ik)*th entry of* $\mathbf{N}^{-\frac{1}{2}}$ *for all* $i, k = 1, \ldots, n$.

(ii) If $\operatorname{rank}\mathbf{A} = n$*, then*

$$\tilde{x}_i = \frac{\operatorname{cdet}_i(\mathbf{A}^*\mathbf{M}\mathbf{A})_{.i}(\mathbf{f})}{\det(\mathbf{A}^*\mathbf{M}\mathbf{A})},$$

where $\mathbf{f} = \mathbf{A}^*\mathbf{M}\mathbf{b} \in \mathbb{H}^{n\times 1}$ *for all* $j = 1, \ldots, m$.

Proof. The proof is similar to the proof of Theorem 5.2 using component-wise representations of $\mathbf{A}^\dagger_{M,N}$ by (42) in the (i) point and by (43) in the (ii) point, respectively. □

Consider a left system of linear equations over the quaternion skew field,

$$\mathbf{x}\mathbf{A} = \mathbf{b} \tag{50}$$

where $\mathbf{A} \in \mathbb{H}^{m\times n}$ is a coefficient matrix, $\mathbf{b} \in \mathbb{H}^{1\times n}$ is a row of constants, and $\mathbf{x} \in \mathbb{H}^{1\times m}$ is a unknown row. The following theorem characterizes the weighted Moore-Penrose solution of (50).

Theorem 5.4. *The left system of linear equations (50) with restriction* $\mathbf{x} \in \mathcal{R}_l(\mathbf{A}^\sharp)$ *has the unique solution* $\tilde{\mathbf{x}} = \mathbf{b}\mathbf{A}^\dagger_{M,N}$.

Theorem 5.5. *Let $\mathbf{A} \in \mathbb{H}^{m\times n}$ and $\mathbf{A}\mathbf{A}^\sharp \in \mathbb{H}^{m\times m}$ be Hermitian.*

(i) If $\operatorname{rank}\mathbf{A} = r \le n < m$*, then the restricted solution* $\tilde{\mathbf{x}} = (\tilde{x}_1, \ldots, \tilde{x}_m)$ *of (50) possess the following determinantal representation*

$$\tilde{x}_j = \frac{\sum_{\alpha\in I_{r,\,m}\{j\}} \operatorname{rdet}_j\left(\left(\mathbf{A}\mathbf{A}^\sharp\right)_{j.}(\mathbf{g})\right)_\alpha^\alpha}{\sum_{\alpha\in I_{r,\,m}} \left|(\mathbf{A}\mathbf{A}^\sharp)_\alpha^\alpha\right|},$$

where $\mathbf{g} = \mathbf{b}\,\mathbf{A}^\sharp \in \mathbb{H}^{1\times m}$.

(ii) If $\operatorname{rank} \mathbf{A} = m$*, then*

$$\tilde{x}_j = \frac{\operatorname{rdet}_j \left(\mathbf{A}\mathbf{A}^\sharp\right)_{j.} (\mathbf{g})}{\det \mathbf{A}\mathbf{A}^\sharp}.$$

Proof. The proof is similar to the proof of Theorem 5.2 using component-wise representations of $\mathbf{A}^\dagger_{M,N}$ by (34) in the (i) point, and (37) in the (ii) point, respectively. □

Theorem 5.6. *Let* $\mathbf{A} \in \mathbb{H}^{m\times n}$ *and* $\mathbf{A}\mathbf{A}^\sharp \in \mathbb{H}^{m\times m}$ *be non-Hermitian.*

(i) If $\operatorname{rank} \mathbf{A} = k \leq n < m$*, then the solution* $\tilde{\mathbf{x}} = (\tilde{x}_1, \dots, \tilde{x}_m)$ *of (50) possess the following determinantal representation*

$$\tilde{x}_j = \frac{\sum\limits_{l} \sum\limits_{\alpha \in I_{r,\,m}\{l\}} \operatorname{rdet}_l \left(\left(\widetilde{\mathbf{A}}\widetilde{\mathbf{A}}^*\right)_{l.} (\mathbf{g}) \right)_{\alpha}^{\alpha} m_{lj}^{\frac{1}{2}}}{\sum\limits_{\alpha \in I_{r,\,m}} \left| \left(\widetilde{\mathbf{A}}\widetilde{\mathbf{A}}^*\right)_{\alpha}^{\alpha} \right|},$$

where $\mathbf{g} = \mathbf{b}\left(\mathbf{N}^{-1}\mathbf{A}^\sharp\mathbf{M}^{\frac{1}{2}}\right) \in \mathbb{H}^{1\times m}$ *and* $m_{lj}^{(\frac{1}{2})}$ *is the* (lj)*th entry of* $\mathbf{M}^{\frac{1}{2}}$ *for all* $l, j = 1, \dots, m$.

(ii) If $\operatorname{rank} \mathbf{A} = m$*, then*

$$\tilde{x}_j = \frac{\operatorname{rdet}_j (\mathbf{A}\mathbf{N}^{-1}\mathbf{A}^*)_{j.} (\mathbf{g})}{\det(\mathbf{A}\mathbf{N}^{-1}\mathbf{A}^*)}.$$

where $\mathbf{g} = \mathbf{b}\mathbf{N}^{-1}\mathbf{A}^* \in \mathbb{H}^{1\times m}$.

Proof. The proof is similar to the proof of Theorem 5.2 using component-wise representations of $\mathbf{A}^\dagger_{M,N}$ by (44) in the (i) point, and (45) in the (ii) point, respectively. □

6. Cramer's Rule for Two-sided Restricted Quaternion Matrix Equation

Definition 6.1. *For an arbitrary matrix over the quaternion skew field,* $\mathbf{A} \in \mathbb{H}^{m\times n}$*, we denote by*

- $\mathcal{R}_r(\mathbf{A}) = \{\mathbf{y} \in \mathbb{H}^{m\times 1} : \mathbf{y} = \mathbf{A}\mathbf{x}, \mathbf{x} \in \mathbb{H}^{n\times 1}\}$, *the column right space of* $\mathbf{A}$,
- $\mathcal{N}_r(\mathbf{A}) = \{\mathbf{x} \in \mathbb{H}^{n\times 1} : \ \mathbf{A}\mathbf{x} = 0\}$, *the right null space of* $\mathbf{A}$,
- $\mathcal{R}_l(\mathbf{A}) = \{\mathbf{y} \in \mathbb{H}^{1\times n} : \ \mathbf{y} = \mathbf{x}\mathbf{A}, \ \mathbf{x} \in \mathbb{H}^{1\times m}\}$, *the row left space of* $\mathbf{A}$,
- $\mathcal{N}_l(\mathbf{A}) = \{\mathbf{x} \in \mathbb{H}^{1\times m} : \ \mathbf{x}\mathbf{A} = 0\}$, *the left null space of* $\mathbf{A}$.

It is easy to see, if $\mathbf{A} \in \mathbb{H}^{n\times n}_n$, then $\mathcal{R}_r \oplus \mathcal{N}_r = \mathbb{H}^{n\times 1}$, and $\mathcal{R}_l \oplus \mathcal{N}_l = \mathbb{H}^{1\times n}$. Suppose that $\mathbf{A} \in \mathbb{H}^{m\times n}$, $\mathbf{B} \in \mathbb{H}^{p\times q}$. Denote

$$\begin{aligned}
\mathcal{R}_r(\mathbf{A},\mathbf{B}) &:= \mathcal{N}_r(\mathbf{Y}) = \{\mathbf{Y} = \mathbf{A}\mathbf{X}\mathbf{B} : \mathbf{X}^{n\times q}\},\\
\mathcal{N}_r(\mathbf{A},\mathbf{B}) &:= \mathcal{R}_r(\mathbf{X}) = \{\mathbf{X}^{n\times p} : \mathbf{A}\mathbf{X}\mathbf{B} = \mathbf{0}\},\\
\mathcal{R}_l(\mathbf{A},\mathbf{A} :) &= \mathcal{R}_l(\mathbf{Y}) = \{\mathbf{Y} = \mathbf{A}\mathbf{X}\mathbf{B} : \mathbf{X}^{n\times q}\},\\
\mathcal{N}_l(\mathbf{A},\mathbf{B}) &:= \mathcal{N}_l(\mathbf{X}) = \{\mathbf{X}^{n\times p} : \mathbf{A}\mathbf{X}\mathbf{B} = \mathbf{0}\}.
\end{aligned}$$

Lemma 6.1. *[28] Suppose that* $\mathbf{A} \in \mathbb{H}^{m\times n}_{r_1}$, $\mathbf{B} \in \mathbb{H}^{p\times q}_{r_2}$, $\mathbf{M}$, $\mathbf{N}$, $\mathbf{P}$, *and* $\mathbf{Q}$ *are Hermitian positive definite matrices of order* m, n, p, *and* q, *respectively. Denote* $\mathbf{A}^\sharp = \mathbf{N}^{-1}\mathbf{A}^*\mathbf{M}$ *and* $\mathbf{B}^\sharp = \mathbf{Q}^{-1}\mathbf{B}^*\mathbf{P}$. *If* $\mathbf{D} \subset \mathcal{R}_r\left(\mathbf{A}\mathbf{A}^\sharp, \mathbf{B}^\sharp\mathbf{B}\right)$ *and* $\mathbf{D} \subset \mathcal{R}_l\left(\mathbf{A}^\sharp\mathbf{A}, \mathbf{B}\mathbf{B}^\sharp\right)$,

$$\mathbf{A}\mathbf{X}\mathbf{B} = \mathbf{D}, \tag{51}$$

$$\mathcal{R}_r(\mathbf{X}) \subset \mathbf{N}^{-1}\mathcal{R}_r(\mathbf{A}^*),\ \mathcal{N}_r(\mathbf{X}) \supset \mathbf{P}^{-1}\mathcal{N}_r(\mathbf{B}^*), \tag{52}$$

$$\mathcal{R}_l(\mathbf{X}) \subset \mathcal{R}_l(\mathbf{A}^*)\mathbf{M},\ \mathcal{N}_l(\mathbf{X}) \supset \mathcal{N}_l(\mathbf{B}^*)\mathbf{Q} \tag{53}$$

then the unique solution of (51) with the restrictions (52)-(53) is

$$\mathbf{X} = \mathbf{A}^\dagger_{M,N}\mathbf{D}\mathbf{B}^\dagger_{P,Q}. \tag{54}$$

In this chapter, we get determinantal representations of (54) that are intrinsically analogs of the classical Cramer's rule. We will consider several cases depending on whether the matrices $\mathbf{A}^\sharp\mathbf{A}$ and $\mathbf{B}\mathbf{B}^\sharp$ are Hermitian or not.

6.1. The Case of Both Hermitian Matrices $\mathbf{A}^\sharp\mathbf{A}$ and $\mathbf{B}\mathbf{B}^\sharp$

Denote $\widetilde{\mathbf{D}} = \mathbf{A}^\sharp\mathbf{D}\mathbf{B}^\sharp$.

Theorem 6.1. *Let* $\mathbf{A}^\sharp\mathbf{A}$ *and* $\mathbf{B}\mathbf{B}^\sharp$ *be Hermitian. Then the solution (54) possess the following determinantal representations.*

(i) If $\operatorname{rank}\mathbf{A} = r_1 < n$ *and* $\operatorname{rank}\mathbf{B} = r_2 < p$*, then*

$$x_{ij} = \frac{\sum\limits_{\beta \in J_{r_1,\,n}\{i\}} \operatorname{cdet}_i \left(\left(\mathbf{A}^\sharp \mathbf{A}\right)_{.\,i} \left(\mathbf{d}^{\,\mathbf{B}}_{\,.j}\right)\right)^{\beta}_{\beta}}{\sum\limits_{\beta \in J_{r_1,n}} \left| \left(\mathbf{A}^\sharp \mathbf{A}\right)^{\beta}_{\beta} \right| \sum\limits_{\alpha \in I_{r_2,p}} \left| \left(\mathbf{B}\mathbf{B}^\sharp\right)^{\alpha}_{\alpha} \right|}, \tag{55}$$

or

$$x_{ij} = \frac{\sum\limits_{\alpha \in I_{r_2,p}\{j\}} \operatorname{rdet}_j \left(\left(\mathbf{B}\mathbf{B}^\sharp\right)_{j\,.} \left(\mathbf{d}^{\,\mathbf{A}}_{\,i\,.}\right)\right)^{\alpha}_{\alpha}}{\sum\limits_{\beta \in J_{r_1,n}} \left| \left(\mathbf{A}^\sharp \mathbf{A}\right)^{\beta}_{\beta} \right| \sum\limits_{\alpha \in I_{r_2,p}} \left| \left(\mathbf{B}\mathbf{B}^\sharp\right)^{\alpha}_{\alpha} \right|}, \tag{56}$$

where

$$\mathbf{d}^{\mathbf{B}}_{.j} = \left(\sum_{\alpha \in I_{r_2,p}\{j\}} \operatorname{rdet}_j \left(\left(\mathbf{B}\mathbf{B}^\sharp\right)_{j.} \left(\tilde{\mathbf{d}}_{k.}\right)\right)^{\alpha}_{\alpha} \right) \in \mathbb{H}^{n\times 1} \tag{57}$$

$$\mathbf{d}^{\mathbf{A}}_{i\,.} = \left(\sum_{\beta \in J_{r_1,n}\{i\}} \operatorname{cdet}_i \left(\left(\mathbf{A}^\sharp \mathbf{A}\right)_{.i} \left(\tilde{\mathbf{d}}_{.l}\right)\right)^{\beta}_{\beta} \right) \in \mathbb{H}^{1\times p} \tag{58}$$

are the column-vector and the row-vector, respectively. $\tilde{\mathbf{d}}_{k.}$ *and* $\tilde{\mathbf{d}}_{.l}$ *are the kth row and the lth column of* $\widetilde{\mathbf{D}}$ *for all* $k = 1, ..., n$*,* $l = 1, ..., p$*.*

(ii) If $\operatorname{rank}\mathbf{A} = n$ *and* $\operatorname{rank}\mathbf{B} = p$*, then*

$$x_{i\,j} = \frac{\operatorname{cdet}_i (\mathbf{A}^\sharp \mathbf{A})_{.\,i} \left(\mathbf{d}^{\mathbf{B}}_{.j}\right)}{\det(\mathbf{A}^\sharp \mathbf{A}) \cdot \det(\mathbf{B}\mathbf{B}^\sharp)}, \tag{59}$$

or

$$x_{i\,j} = \frac{\operatorname{rdet}_j (\mathbf{B}\mathbf{B}^\sharp)_{j.} \left(\mathbf{d}^{\mathbf{A}}_{i\,.}\right)}{\det(\mathbf{A}^\sharp \mathbf{A}) \cdot \det(\mathbf{B}\mathbf{B}^\sharp)}, \tag{60}$$

where

$$\mathbf{d}^{\mathbf{B}}_{.j} := \left(\operatorname{rdet}_j (\mathbf{B}\mathbf{B}^\sharp)_{j.} \left(\tilde{\mathbf{d}}_{k\,.}\right)\right) \in \mathbb{H}^{n\times 1}, \tag{61}$$

$$\mathbf{d}^{\mathbf{A}}_{i\,.} := \left(\operatorname{cdet}_i (\mathbf{A}^\sharp \mathbf{A})_{.\,i} \left(\tilde{\mathbf{d}}_{.l}\right)\right) \in \mathbb{H}^{1\times p}. \tag{62}$$

(iii) *If* $\operatorname{rank}\mathbf{A} = n$ *and* $\operatorname{rank}\mathbf{B} = r_2 < p$, *then*

$$x_{ij} = \frac{\operatorname{cdet}_i\left(\left(\mathbf{A}^\sharp\mathbf{A}\right)_{.i}\left(\mathbf{d}^{\mathbf{B}}_{.j}\right)\right)}{\det(\mathbf{A}^\sharp\mathbf{A}) \cdot \sum\limits_{\alpha\in I_{r_2,p}} \left|(\mathbf{B}\mathbf{B}^\sharp)^\alpha_\alpha\right|}, \tag{63}$$

or

$$x_{ij} = \frac{\sum\limits_{\alpha\in I_{r_2,p}\{j\}} \operatorname{rdet}_j\left(\left(\mathbf{B}\mathbf{B}^\sharp\right)_{j.}\left(\mathbf{d}^{\mathbf{A}}_{i.}\right)\right)^\alpha_\alpha}{\det(\mathbf{A}^\sharp\mathbf{A}) \cdot \sum\limits_{\alpha\in I_{r_2,p}} \left|(\mathbf{B}\mathbf{B}^\sharp)^\alpha_\alpha\right|}, \tag{64}$$

where $\mathbf{d}^{\mathbf{B}}_{.j}$ *is (57) and* $\mathbf{d}^{\mathbf{A}}_{i.}$ *is (62).*

(iv) *If* $\operatorname{rank}\mathbf{A} = r_1 < n$ *and* $\operatorname{rank}\mathbf{B} = p$, *then*

$$x_{ij} = \frac{\operatorname{rdet}_j(\mathbf{B}\mathbf{B}^\sharp)_{j.}\left(\mathbf{d}^{\mathbf{A}}_{i.}\right)}{\sum\limits_{\beta\in J_{r_1,n}} \left|(\mathbf{A}^\sharp\mathbf{A})^\beta_\beta\right| \cdot \det(\mathbf{B}\mathbf{B}^\sharp)}, \tag{65}$$

or

$$x_{ij} = \frac{\sum\limits_{\beta\in J_{r_1,n}\{i\}} \operatorname{cdet}_i\left(\left(\mathbf{A}^\sharp\mathbf{A}\right)_{.i}\left(\mathbf{d}^{\mathbf{B}}_{.j}\right)\right)^\beta_\beta}{\sum\limits_{\beta\in J_{r_1,n}} \left|(\mathbf{A}^\sharp\mathbf{A})^\beta_\beta\right| \cdot \det(\mathbf{B}\mathbf{B}^\sharp)}, \tag{66}$$

where $\mathbf{d}^{\mathbf{B}}_{.j}$ *is (61) and* $\mathbf{d}^{\mathbf{A}}_{i.}$ *is (58).*

Proof. (i) If $\mathbf{A} \in \mathbb{H}^{m\times n}_{r_1}$, $\mathbf{B} \in \mathbb{H}^{p\times q}_{r_2}$ and $r_1 < n$, $r_2 < p$, then, by Theorem 4.1, the weighted Moore-Penrose inverses $\mathbf{A}^\dagger = \left(a^\ddagger_{ij}\right) \in \mathbb{H}^{n\times m}$ and $\mathbf{B}^\dagger = \left(b^\ddagger_{ij}\right) \in \mathbb{H}^{q\times p}$ possess the following determinantal representations, respectively,

$$a^\ddagger_{ij} = \frac{\sum\limits_{\beta\in J_{r_1,n}\{i\}} \operatorname{cdet}_i\left(\left(\mathbf{A}^\sharp\mathbf{A}\right)_{.i}\left(\mathbf{a}^\sharp_{.j}\right)\right)^\beta_\beta}{\sum\limits_{\beta\in J_{r_1,n}} \left|\left(\mathbf{A}^\sharp\mathbf{A}\right)^\beta_\beta\right|}, \tag{67}$$

$$b^\ddagger_{ij} = \frac{\sum\limits_{\alpha\in I_{r_2,p}\{j\}} \operatorname{rdet}_j\left((\mathbf{B}\mathbf{B}^\sharp)_{j.}(\mathbf{b}^\sharp_{i.})\right)^\alpha_\alpha}{\sum\limits_{\alpha\in I_{r_2,p}} \left|(\mathbf{B}\mathbf{B}^\sharp)^\alpha_\alpha\right|}. \tag{68}$$

By Lemma 6.1, $\mathbf{X} = \mathbf{A}^{\dagger}_{M,N}\mathbf{D}\mathbf{B}^{\dagger}_{P,Q}$ and entries of $\mathbf{X} = (x_{ij})$ are

$$x_{ij} = \sum_{s=1}^{q}\left(\sum_{k=1}^{m} a^{\ddagger}_{ik} d_{ks}\right) b^{\ddagger}_{sj}. \tag{69}$$

for all $i = 1, ..., n, j = 1, ..., p$.

Denote by $\hat{\mathbf{d}}_{.s}$ the sth column of $\mathbf{A}^{\sharp}\mathbf{D} =: \hat{\mathbf{D}} = (\hat{d}_{ij}) \in \mathbb{H}^{n\times q}$ for all $s = 1, ..., q$. It follows from $\sum\limits_{k} \mathbf{a}^{\sharp}_{.\,k} d_{ks} = \hat{\mathbf{d}}_{.\,s}$ that

$$\sum_{k=1}^{m} a^{\ddagger}_{ik} d_{ks} = \sum_{k=1}^{m} \frac{\sum\limits_{\beta\in J_{r_1,\,n}\{i\}} \mathrm{cdet}_i \left(\left(\mathbf{A}^{\sharp}\mathbf{A}\right)_{.\,i}\left(\mathbf{a}^{\sharp}_{.k}\right)\right)^{\beta}_{\beta}}{\sum\limits_{\beta\in J_{r_1,\,n}} \left|\left(\mathbf{A}^{\sharp}\mathbf{A}\right)^{\beta}_{\beta}\right|} \cdot d_{ks} =$$

$$\frac{\sum\limits_{\beta\in J_{r_1,\,n}\{i\}} \sum\limits_{k=1}^{m} \mathrm{cdet}_i \left(\left(\mathbf{A}^{\sharp}\mathbf{A}\right)_{.\,i}\left(\mathbf{a}^{\sharp}_{.k}\right)\right)^{\beta}_{\beta} \cdot d_{ks}}{\sum\limits_{\beta\in J_{r_1,\,n}} \left|\left(\mathbf{A}^{\sharp}\mathbf{A}\right)^{\beta}_{\beta}\right|} = \frac{\sum\limits_{\beta\in J_{r_1,\,n}\{i\}} \mathrm{cdet}_i \left(\left(\mathbf{A}^{\sharp}\mathbf{A}\right)_{.\,i}\left(\hat{\mathbf{d}}_{.\,s}\right)\right)^{\beta}_{\beta}}{\sum\limits_{\beta\in J_{r_1,\,n}} \left|\left(\mathbf{A}^{\sharp}\mathbf{A}\right)^{\beta}_{\beta}\right|}. \tag{70}$$

Suppose $\mathbf{e}_{s.}$ and $\mathbf{e}_{.\,s}$ are the unit row-vector and the unit column-vector, respectively, such that all their components are 0, except the sth components, which are 1. Substituting (70) and (68) in (69), we obtain

$$x_{ij} = \sum_{s=1}^{q} \frac{\sum\limits_{\beta\in J_{r_1,\,n}\{i\}} \mathrm{cdet}_i \left(\left(\mathbf{A}^{\sharp}\mathbf{A}\right)_{.\,i}\left(\hat{\mathbf{d}}_{.\,s}\right)\right)^{\beta}_{\beta}}{\sum\limits_{\beta\in J_{r_1,\,n}} \left|\left(\mathbf{A}^{\sharp}\mathbf{A}\right)^{\beta}_{\beta}\right|} \frac{\sum\limits_{\alpha\in I_{r_2,p}\{j\}} \mathrm{rdet}_j \left((\mathbf{B}\mathbf{B}^{\sharp})_{j\,.}(\mathbf{b}^{\sharp}_{s.})\right)^{\alpha}_{\alpha}}{\sum\limits_{\alpha\in I_{r_2,p}} \left|(\mathbf{B}\mathbf{B}^{\sharp})^{\alpha}_{\alpha}\right|}.$$

Since

$$\hat{\mathbf{d}}_{.\,s} = \sum_{l=1}^{n} \mathbf{e}_{.\,l}\hat{d}_{ls},\ \mathbf{b}^{\sharp}_{s.} = \sum_{t=1}^{p} b^{\sharp}_{st}\mathbf{e}_{t.},\ \sum_{s=1}^{q} \hat{d}_{ls} b^{\sharp}_{st} = \widetilde{d}_{lt},$$

then we have

$$x_{ij} =$$

$$\frac{\sum\limits_{s=1}^{q}\sum\limits_{t=1}^{p}\sum\limits_{l=1}^{n}\sum\limits_{\beta\in J_{r_1,\,n}\{i\}} \mathrm{cdet}_i \left(\left(\mathbf{A}^{\sharp}\mathbf{A}\right)_{.\,i}(\mathbf{e}_{.\,l})\right)^{\beta}_{\beta} \hat{d}_{ls} b^{\sharp}_{st} \sum\limits_{\alpha\in I_{r_2,p}\{j\}} \mathrm{rdet}_j \left((\mathbf{B}\mathbf{B}^{\sharp})_{j\,.}(\mathbf{e}_{t.})\right)^{\alpha}_{\alpha}}{\sum\limits_{\beta\in J_{r_1,\,n}} \left|\left(\mathbf{A}^{\sharp}\mathbf{A}\right)^{\beta}_{\beta}\right| \sum\limits_{\alpha\in I_{r_2,p}} \left|(\mathbf{B}\mathbf{B}^{\sharp})^{\alpha}_{\alpha}\right|} =$$

$$\frac{\sum\limits_{t=1}^{p}\sum\limits_{l=1}^{n}\sum\limits_{\beta\in J_{r_1,\,n}\{i\}}\mathrm{cdet}_i\left(\left(\mathbf{A}^{\sharp}\mathbf{A}\right)_{.\,i}(\mathbf{e}_{.\,l})\right)_{\beta}^{\beta}\widetilde{d}_{lt}\sum\limits_{\alpha\in I_{r_2,p}\{j\}}\mathrm{rdet}_j\left((\mathbf{B}\mathbf{B}^{\sharp})_{j\,.}(\mathbf{e}_{t.})\right)_{\alpha}^{\alpha}}{\sum\limits_{\beta\in J_{r_1,\,n}}\left|\left(\mathbf{A}^{\sharp}\mathbf{A}\right)_{\beta}^{\beta}\right|\sum\limits_{\alpha\in I_{r_2,p}}\left|(\mathbf{B}\mathbf{B}^{\sharp})_{\alpha}^{\alpha}\right|}. \tag{71}$$

Denote by

$$d_{it}^{\mathbf{A}} :=$$

$$\sum_{\beta\in J_{r_1,\,n}\{i\}}\mathrm{cdet}_i\left(\left(\mathbf{A}^{\sharp}\mathbf{A}\right)_{.\,i}\left(\widetilde{\mathbf{d}}_{.t}\right)\right)_{\beta}^{\beta}=\sum_{l=1}^{n}\sum_{\beta\in J_{r_1,n}\{i\}}\mathrm{cdet}_i\left(\left(\mathbf{A}^{\sharp}\mathbf{A}\right)_{.\,i}(\mathbf{e}_{.l})\right)_{\beta}^{\beta}\widetilde{d}_{lt}$$

the tth component of a row-vector $\mathbf{d}_{i\,.}^{\mathbf{A}} = (d_{i1}^{\mathbf{A}}, ..., d_{ip}^{\mathbf{A}})$ for all $t = 1, ..., p$. Substituting it in (71), we have

$$x_{ij} = \frac{\sum\limits_{t=1}^{p} d_{it}^{\mathbf{A}}\sum\limits_{\alpha\in I_{r_2,p}\{j\}}\mathrm{rdet}_j\left((\mathbf{B}\mathbf{B}^{\sharp})_{j\,.}(\mathbf{e}_{t.})\right)_{\alpha}^{\alpha}}{\sum\limits_{\beta\in J_{r_1,\,n}}\left|\left(\mathbf{A}^{\sharp}\mathbf{A}\right)_{\beta}^{\beta}\right|\sum\limits_{\alpha\in I_{r_2,p}}\left|(\mathbf{B}\mathbf{B}^{\sharp})_{\alpha}^{\alpha}\right|}.$$

Since $\sum\limits_{t=1}^{p} d_{it}^{\mathbf{A}}\mathbf{e}_{t.} = \mathbf{d}_{i\,.}^{\mathbf{A}}$, then it follows (56).

If we denote by

$$d_{lj}^{\mathbf{B}} := \sum_{t=1}^{p}\widetilde{d}_{lt}\sum_{\alpha\in I_{r_2,p}\{j\}}\mathrm{rdet}_j\left((\mathbf{B}\mathbf{B}^{\sharp})_{j\,.}(\mathbf{e}_{t.})\right)_{\alpha}^{\alpha} = \sum_{\alpha\in I_{r_2,p}\{j\}}\mathrm{rdet}_j\left((\mathbf{B}\mathbf{B}^{\sharp})_{j\,.}(\widetilde{\mathbf{d}}_{l.})\right)_{\alpha}^{\alpha}$$

the lth component of a column-vector $\mathbf{d}_{.\,j}^{\mathbf{B}} = (d_{1j}^{\mathbf{B}}, ..., d_{jn}^{\mathbf{B}})^T$ for all $l = 1, ..., n$ and substitute it in (71), we obtain

$$x_{ij} = \frac{\sum\limits_{l=1}^{n}\sum\limits_{\beta\in J_{r_1,\,n}\{i\}}\mathrm{cdet}_i\left(\left(\mathbf{A}^{\sharp}\mathbf{A}\right)_{.\,i}(\mathbf{e}_{.\,l})\right)_{\beta}^{\beta} d_{lj}^{\mathbf{B}}}{\sum\limits_{\beta\in J_{r_1,\,n}}\left|\left(\mathbf{A}^{\sharp}\mathbf{A}\right)_{\beta}^{\beta}\right|\sum\limits_{\alpha\in I_{r_2,p}}\left|(\mathbf{B}\mathbf{B}^{\sharp})_{\alpha}^{\alpha}\right|}.$$

Since $\sum\limits_{l=1}^{n}\mathbf{e}_{.l}d_{lj}^{\mathbf{B}} = \mathbf{d}_{.\,j}^{\mathbf{B}}$, then it follows (55).

(ii) If $\operatorname{rank} \mathbf{A} = n$ and $\operatorname{rank} \mathbf{B} = p$, then by Theorem 4.1 the weighted Moore-Penrose inverses $\mathbf{A}^{\dagger}_{M,N} = \left(a^{\ddagger}_{ij}\right) \in \mathbb{H}^{n\times m}$ and $\mathbf{B}^{\dagger}_{P,Q} = \left(b^{\ddagger}_{ij}\right) \in \mathbb{H}^{q\times p}$ possess the following determinantal representations, respectively,

$$a^{\ddagger}_{ij} = \frac{\operatorname{cdet}_i(\mathbf{A}^{\sharp}\mathbf{A})_{.\,i}\left(\mathbf{a}^{\sharp}_{.\,j}\right)}{\det(\mathbf{A}^{\sharp}\mathbf{A})} \tag{72}$$

$$b^{\ddagger}_{ij} = \frac{\operatorname{rdet}_j(\mathbf{B}\mathbf{B}^{\sharp})_{j.}\left(\mathbf{b}^{\sharp}_{i.}\right)}{\det(\mathbf{B}\mathbf{B}^{\sharp})}. \tag{73}$$

By their substituting in (69) and pondering ahead as in the previous case, we obtain (59) and (60).

(iii) If $\mathbf{A} \in \mathbb{H}^{m\times n}_{r_1}$, $\mathbf{B} \in \mathbb{H}^{p\times q}_{r_2}$ and $r_1 = n$, $r_2 < p$, then, for the weighted Moore-Penrose inverses $\mathbf{A}^{\dagger}_{M,N}$ and $\mathbf{B}^{\dagger}_{P,Q}$, the determinantal representations (72) and (67) are more applicable to use, respectively. By their substituting in (69) and pondering ahead as in the previous case, we finally obtain (63) and (64) as well.

(iv) In this case for $\mathbf{A}^{\dagger}_{M,N}$ and $\mathbf{B}^{\dagger}_{P,Q}$, we use the determinantal representations (72) and (68), respectively. □

Corollary 6.1. *Suppose that* $\mathbf{A} \in \mathbb{H}^{m\times n}_{r_1}$, $\mathbf{D} \in \mathbb{H}^{m\times p}$, $\mathbf{M}$, $\mathbf{N}$ *are Hermitian positive definite matrices of order* m *and* n, *respectively,* $\mathbf{A}^{\sharp}\mathbf{A}$ *is Hermitian. Denote* $\widehat{\mathbf{D}} = \mathbf{A}^{\sharp}\mathbf{D}$. *If* $\mathbf{D} \subset \mathcal{R}_r(\mathbf{A}\mathbf{A}^{\sharp})$ *and* $\mathbf{D} \subset \mathcal{R}_l(\mathbf{A}^{\sharp}\mathbf{A})$,

$$\mathbf{A}\mathbf{X} = \mathbf{D}, \tag{74}$$

$$\mathcal{R}_r(\mathbf{X}) \subset \mathbf{N}^{-1}\mathcal{R}_r(\mathbf{A}^*),\ \mathcal{R}_l(\mathbf{X}) \subset \mathcal{R}_l(\mathbf{A}^*)\mathbf{M}, \tag{75}$$

then the unique solution of (74) with the restrictions (75) is

$$\mathbf{X} = \mathbf{A}^{\dagger}_{M,N}\mathbf{D}$$

which possess the following determinantal representations.

(i) If $\operatorname{rank} \mathbf{A} = r_1 < n$, *then*

$$x_{ij} = \frac{\sum\limits_{\beta\in J_{r_1,\,n}\{i\}} \operatorname{cdet}_i\left(\left(\mathbf{A}^{\sharp}\mathbf{A}\right)_{.\,i}\left(\widehat{\mathbf{d}}_{.\,j}\right)\right)^{\beta}_{\beta}}{\sum\limits_{\beta\in J_{r_1,n}} \left|\left(\mathbf{A}^{\sharp}\mathbf{A}\right)^{\beta}_{\beta}\right|},$$

where $\widehat{\mathbf{d}}_{.j}$ *are the* j*th column of* $\widehat{\mathbf{D}}$ *for all* $i = 1, ..., n$, $j = 1, ..., p$.

(ii) If $\operatorname{rank}\mathbf{A} = n$, *then*

$$x_{i\,j} = \frac{\operatorname{cdet}_i(\mathbf{A}^\sharp\mathbf{A})_{.\,i}\left(\widehat{\mathbf{d}}_{.j}\right)}{\det(\mathbf{A}^\sharp\mathbf{A})},$$

Proof. The proof follows evidently from Theorem 6.1 when $\mathbf{B}$ be removed, and unit matrices insert instead $\mathbf{P}$, $\mathbf{Q}$.

Corollary 6.2. *Suppose that* $\mathbf{B} \in \mathbb{H}^{p\times q}_{r_2}$, $\mathbf{D} \in \mathbb{H}^{n\times q}$, $\mathbf{P}$, *and* $\mathbf{Q}$ *are Hermitian positive definite matrices of order* p *and* q, *respectively,* $\mathbf{B}\mathbf{B}^\sharp$ *is Hermitian. Denote* $\check{\mathbf{D}} = \mathbf{D}\mathbf{B}^\sharp$. *If* $\mathbf{D} \subset \mathcal{R}_r(\mathbf{B}^\sharp\mathbf{B})$ *and* $\mathbf{D} \subset \mathcal{R}_l(\mathbf{B}\mathbf{B}^\sharp)$,

$$\mathbf{X}\mathbf{B} = \mathbf{D}, \tag{76}$$

$$\mathcal{N}_r(\mathbf{X}) \supset \mathbf{P}^{-1}\mathcal{N}_r(\mathbf{B}^*),\, \mathcal{N}_l(\mathbf{X}) \supset \mathcal{N}_l(\mathbf{B}^*)\mathbf{Q}, \tag{77}$$

then the unique solution of (76) with the restrictions (77) is

$$\mathbf{X} = \mathbf{D}\mathbf{B}^\dagger_{P,Q}$$

which possess the following determinantal representations.

(i) If $\operatorname{rank}\mathbf{B} = r_2 < p$, *then*

$$x_{ij} = \frac{\sum\limits_{\alpha\in I_{r_2,q}\{j\}} \operatorname{rdet}_j\left(\left(\mathbf{B}\mathbf{B}^\sharp\right)_{j.}\left(\check{\mathbf{d}}_{\,i.}\right)\right)\,{}^{\alpha}_{\alpha}}{\sum\limits_{\alpha\in I_{r_2,q}} \left|\left(\mathbf{B}\mathbf{B}^\sharp\right)^{\alpha}_{\alpha}\right|},$$

where $\check{\mathbf{d}}_{i.}$ *are the ith row of* $\check{\mathbf{D}}$ *for all* $i = 1, ..., n$, $j = 1, ..., p$.

(ii) If $\operatorname{rank}\mathbf{B} = p$, *then*

$$x_{i\,j} = \frac{\operatorname{rdet}_j\left(\mathbf{B}\mathbf{B}^\sharp\right)_{j.}\left(\check{\mathbf{d}}_{i.}\right)}{\det\left(\mathbf{B}\mathbf{B}^\sharp\right)}.$$

Proof. The proof follows evidently from Theorem 6.1 when $\mathbf{A}$ be removed and unit matrices insert instead $\mathbf{M}$, $\mathbf{N}$.

6.2. The Case of Both Non-Hermitian Matrices $\mathbf{A}^{\sharp}\mathbf{A}$ and $\mathbf{B}\mathbf{B}^{\sharp}$

Theorem 6.2. *Let* $\mathbf{A}^{\sharp}\mathbf{A}$ *and* $\mathbf{B}\mathbf{B}^{\sharp}$ *be both non-Hermitian. Then the solution (54) possess the following determinantal representations.*

(i) If $\operatorname{rank}\mathbf{A} = r_1 < n$ *and* $\operatorname{rank}\mathbf{B} = r_2 < p$*, then*

$$x_{ij} = \frac{\sum\limits_{k} n_{ik}^{(-\frac{1}{2})} \sum\limits_{\beta \in J_{r_1,\,n}\{k\}} \operatorname{cdet}_k \left(\left(\widetilde{\mathbf{A}}^* \widetilde{\mathbf{A}} \right)_{.\,k} \left(\mathbf{d}_{.j}^{\mathbf{B}} \right) \right)_{\beta}^{\beta}}{\sum\limits_{\beta \in J_{r_1,\,n}} \left| \left(\widetilde{\mathbf{A}}^* \widetilde{\mathbf{A}} \right)_{\beta}^{\beta} \right| \sum\limits_{\alpha \in I_{r_2,p}} \left| \left(\widetilde{\mathbf{B}} \widetilde{\mathbf{B}}^* \right)_{\alpha}^{\alpha} \right|}, \tag{78}$$

or

$$x_{ij} = \frac{\sum\limits_{l} \sum\limits_{\alpha \in I_{r_2,\,p}\{l\}} \operatorname{rdet}_l \left(\left(\widetilde{\mathbf{B}} \widetilde{\mathbf{B}}^* \right)_{l.} \left(\mathbf{d}_{i.}^{\mathbf{A}} \right) \right)_{\alpha}^{\alpha} \cdot m_{lj}^{(\frac{1}{2})}}{\sum\limits_{\beta \in J_{r_1,\,n}} \left| \left(\widetilde{\mathbf{A}}^* \widetilde{\mathbf{A}} \right)_{\beta}^{\beta} \right| \sum\limits_{\alpha \in I_{r_2,p}} \left| \left(\widetilde{\mathbf{B}} \widetilde{\mathbf{B}}^* \right)_{\alpha}^{\alpha} \right|}, \tag{79}$$

where

$$\mathbf{d}_{.j}^{\mathbf{B}} = \left(\sum\limits_{l} \sum\limits_{\alpha \in I_{r_2,\,p}\{l\}} \operatorname{rdet}_l \left(\left(\widetilde{\mathbf{B}} \widetilde{\mathbf{B}}^* \right)_{l.} \left(\widetilde{\mathbf{d}}_{t.} \right) \right)_{\alpha}^{\alpha} \cdot m_{lj}^{(\frac{1}{2})} \right) \in \mathbb{H}^{n \times 1} \tag{80}$$

$$\mathbf{d}_{i.}^{\mathbf{A}} = \left(\sum\limits_{k} n_{ik}^{(-\frac{1}{2})} \sum\limits_{\beta \in J_{r_1,\,n}\{k\}} \operatorname{cdet}_k \left(\left(\widetilde{\mathbf{A}}^* \widetilde{\mathbf{A}} \right)_{.\,k} \left(\widetilde{\mathbf{d}}_{.f} \right) \right)_{\beta}^{\beta} \right) \in \mathbb{H}^{1 \times p} \tag{81}$$

are the column-vector and the row-vector, respectively. $\widetilde{\mathbf{d}}_{t.}$ *and* $\widetilde{\mathbf{d}}_{.f}$ *are the* t*th row and the* f*th column of* $\widetilde{\mathbf{D}} := \mathbf{N}^{-\frac{1}{2}}\mathbf{A}^*\mathbf{M}\mathbf{D}\mathbf{Q}^{-1}\mathbf{B}^*\mathbf{P}^{\frac{1}{2}} = (\widetilde{d}_{ij}) \in \mathbb{H}^{n \times p}$ *for all* $t = 1, ..., n$, $f = 1, ..., p$.

(ii) If $\operatorname{rank}\mathbf{A} = n$ *and* $\operatorname{rank}\mathbf{B} = p$*, then*

$$x_{ij} = \frac{\operatorname{cdet}_i (\mathbf{A}^*\mathbf{M}\mathbf{A})_{.i} \left(\mathbf{d}_{.j}^{\mathbf{B}} \right)}{\det(\mathbf{A}^*\mathbf{M}\mathbf{A}) \cdot \det(\mathbf{B}\mathbf{Q}^{-1}\mathbf{B}^*)}, \tag{82}$$

or

$$x_{ij} = \frac{\operatorname{rdet}_j (\mathbf{B}\mathbf{Q}^{-1}\mathbf{B}^*)_{j.} \left(\mathbf{d}_{i.}^{\mathbf{A}} \right)}{\det(\mathbf{A}^*\mathbf{M}\mathbf{A}) \cdot \det(\mathbf{B}\mathbf{Q}^{-1}\mathbf{B}^*)}, \tag{83}$$

where

$$\mathbf{d}^{\mathbf{B}}_{.j} := \left(\mathrm{rdet}_j(\mathbf{B}\mathbf{Q}^{-1}\mathbf{B}^*)_{j.}\left(\widetilde{\mathbf{d}}_{t.}\right)\right) \in \mathbb{H}^{n\times 1}, \tag{84}$$

$$\mathbf{d}^{\mathbf{A}}_{i.} := \left(\mathrm{cdet}_i(\mathbf{A}^*\mathbf{M}\mathbf{A})_{.i}\left(\widetilde{\mathbf{d}}_{.f}\right)\right) \in \mathbb{H}^{1\times p}, \tag{85}$$

$\widetilde{\mathbf{d}}_{t.}$ *and* $\widetilde{\mathbf{d}}_{.f}$ *are the* t*th row and the* f*th column of* $\widetilde{\mathbf{D}} := \mathbf{A}^*\mathbf{M}\mathbf{D}\mathbf{Q}^{-1}\mathbf{B}^* \in \mathbb{H}^{n\times p}$*, respectively.*

(iii) If $\operatorname{rank}\mathbf{A} = n$ *and* $\operatorname{rank}\mathbf{B} = r_2 < p$*, then*

$$x_{ij} = \frac{\mathrm{cdet}_i\left((\mathbf{A}^*\mathbf{M}\mathbf{A})_{.i}\left(\mathbf{d}^{\mathbf{B}}_{.j}\right)\right)}{\det(\mathbf{A}^*\mathbf{M}\mathbf{A})\cdot \sum\limits_{\alpha\in I_{r_2,p}} \left|\left(\widetilde{\mathbf{B}}\widetilde{\mathbf{B}}^*\right)^{\alpha}_{\alpha}\right|}, \tag{86}$$

or

$$x_{ij} = \frac{\sum\limits_{l}\sum\limits_{\alpha\in I_{r_2,\,p}\{l\}} \mathrm{rdet}_l\left(\left(\widetilde{\mathbf{B}}\widetilde{\mathbf{B}}^*\right)_{l.}(\mathbf{d}^{\mathbf{A}}_{i.})\right)^{\alpha}_{\alpha}\cdot m^{(\frac{1}{2})}_{lj}}{\det(\mathbf{A}^*\mathbf{M}\mathbf{A})\cdot \sum\limits_{\alpha\in I_{r_2,p}} \left|\left(\widetilde{\mathbf{B}}\widetilde{\mathbf{B}}^*\right)^{\alpha}_{\alpha}\right|}, \tag{87}$$

where $\mathbf{d}^{\mathbf{B}}_{.j}$ *is (80) and* $\mathbf{d}^{\mathbf{A}}_{i.}$ *is (85).*

(iv) If $\operatorname{rank}\mathbf{A} = r_1 < n$ *and* $\operatorname{rank}\mathbf{B} = p$*, then*

$$x_{ij} = \frac{\mathrm{rdet}_j(\mathbf{B}\mathbf{Q}^{-1}\mathbf{B}^*)_{j.}\left(\mathbf{d}^{\mathbf{A}}_{i.}\right)}{\sum\limits_{\beta\in J_{r_1,n}} \left|\left(\widetilde{\mathbf{A}}^*\widetilde{\mathbf{A}}\right)^{\beta}_{\beta}\right|\cdot \det(\mathbf{B}\mathbf{Q}^{-1}\mathbf{B}^*)}, \tag{88}$$

or

$$x_{ij} = \frac{\sum\limits_{k} n^{(-\frac{1}{2})}_{ik} \sum\limits_{\beta\in J_{r_1,\,n}\{k\}} \mathrm{cdet}_k\left(\left(\widetilde{\mathbf{A}}^*\widetilde{\mathbf{A}}\right)_{.k}\left(\mathbf{d}^{\mathbf{B}}_{.j}\right)\right)^{\beta}_{\beta}}{\sum\limits_{\beta\in J_{r_1,n}} \left|\left(\widetilde{\mathbf{A}}^*\widetilde{\mathbf{A}}\right)^{\beta}_{\beta}\right|\cdot \det(\mathbf{B}\mathbf{Q}^{-1}\mathbf{B}^*)}, \tag{89}$$

where $\mathbf{d}^{\mathbf{B}}_{.j}$ *is (84) and* $\mathbf{d}^{\mathbf{A}}_{i.}$ *is (81).*

Proof. (i) If $\mathbf{A} \in \mathbb{H}^{m\times n}_{r_1}$, $\mathbf{B} \in \mathbb{H}^{p\times q}_{r_2}$ are both non-Hermitian, and $r_1 < n$, $r_2 < p$, then, by Theorem 4.2, the weighted Moore-Penrose inverses $\mathbf{A}^\dagger =$

$\left(a^{\ddagger}_{ij}\right) \in \mathbb{H}^{n\times m}$ and $\mathbf{B}^{\dagger} = \left(b^{\ddagger}_{ij}\right) \in \mathbb{H}^{q\times p}$ posses the following determinantal representations, respectively,

$$a^{\ddagger}_{ij} = \frac{\sum\limits_{k} n_{ik}^{(-\frac{1}{2})} \sum\limits_{\beta\in J_{r_1,\,n}\{k\}} \operatorname{cdet}_k \left(\left(\widetilde{\mathbf{A}}^*\widetilde{\mathbf{A}}\right)_{.\,k}(\widehat{\mathbf{a}}_{.j})\right)^{\beta}_{\beta}}{\sum\limits_{\beta\in J_{r_1,\,n}} \left|\left(\widetilde{\mathbf{A}}^*\widetilde{\mathbf{A}}\right)^{\beta}_{\beta}\right|}, \tag{90}$$

where $\widehat{\mathbf{a}}_{.j}$ is the jth column of $\mathbf{N}^{-\frac{1}{2}}\mathbf{A}^*\mathbf{M}$;

$$b^{\ddagger}_{ij} = \frac{\sum\limits_{l} \sum\limits_{\alpha\in I_{r_2,\,p}\{l\}} \operatorname{rdet}_l \left(\left(\widetilde{\mathbf{B}}\widetilde{\mathbf{B}}^*\right)_{l.}(\widehat{\mathbf{b}}_{i.})\right)^{\alpha}_{\alpha} \cdot m_{lj}^{(\frac{1}{2})}}{\sum\limits_{\alpha\in I_{2,\,p}} \left|\left(\widetilde{\mathbf{B}}\widetilde{\mathbf{B}}^*\right)^{\alpha}_{\alpha}\right|}, \tag{91}$$

where $\widehat{\mathbf{b}}_{i.}$ is the ith row of $\mathbf{Q}^{-1}\mathbf{B}^*\mathbf{P}^{\frac{1}{2}}$. By Lemma 6.1, $\mathbf{X} = \mathbf{A}^{\dagger}_{M,N}\mathbf{D}\mathbf{B}^{\dagger}_{P,Q}$ and entries of $\mathbf{X} = (x_{ij})$ are

$$x_{ij} = \sum_{s=1}^{q}\left(\sum_{t=1}^{m} a^{\ddagger}_{it} d_{ts}\right) b^{\ddagger}_{sj}. \tag{92}$$

for all $i = 1, ..., n$, $j = 1, ..., p$.

Denote by $\widehat{\mathbf{d}}_{.s}$ the sth column of $\mathbf{N}^{-\frac{1}{2}}\mathbf{A}^*\mathbf{M}\mathbf{D} =: \widehat{\mathbf{D}} = (\widehat{d}_{ij}) \in \mathbb{H}^{n\times q}$ for all $s = 1, ..., q$. It follows from $\sum\limits_{t} \widehat{\mathbf{a}}_{.\,t} d_{ts} = \widehat{\mathbf{d}}_{.\,s}$ that

$$\sum_{t=1}^{m} a^{\ddagger}_{it} d_{ts} = \sum_{t=1}^{m} \frac{\sum\limits_{k} n_{ik}^{(-\frac{1}{2})} \sum\limits_{\beta\in J_{r_1,\,n}\{i\}} \operatorname{cdet}_k \left(\left(\widetilde{\mathbf{A}}^*\widetilde{\mathbf{A}}\right)_{.\,k}(\widehat{\mathbf{a}}_{.t})\right)^{\beta}_{\beta}}{\sum\limits_{\beta\in J_{r_1,\,n}} \left|\left(\widetilde{\mathbf{A}}^*\widetilde{\mathbf{A}}\right)^{\beta}_{\beta}\right|} \cdot d_{ts} = \frac{\sum\limits_{k} n_{ik}^{(-\frac{1}{2})} \sum\limits_{\beta\in J_{r_1,\,n}\{i\}} \operatorname{cdet}_k \left(\left(\widetilde{\mathbf{A}}^*\widetilde{\mathbf{A}}\right)_{.\,k}\left(\widehat{\mathbf{d}}_{.s}\right)\right)^{\beta}_{\beta}}{\sum\limits_{\beta\in J_{r_1,\,n}} \left|\left(\widetilde{\mathbf{A}}^*\widetilde{\mathbf{A}}\right)^{\beta}_{\beta}\right|}. \tag{93}$$

Suppose $\mathbf{e}_{s.}$ and $\mathbf{e}_{.\,s}$ are the unit row-vector and the unit column-vector, respectively, such that all their components are 0, except the sth components, which are 1. Substituting (93) and (91) in (92), we obtain

$$x_{ij}=\sum_{s=1}^{q}\frac{\sum\limits_{k} n_{ik}^{(-\frac{1}{2})}\sum\limits_{\beta\in J_{r_1,\,n}\{k\}}\text{cdet}_k\left(\left(\widetilde{\mathbf{A}}^*\widetilde{\mathbf{A}}\right)_{.\,k}\left(\widehat{\mathbf{d}}_{.s}\right)\right)_{\beta}^{\beta}}{\sum\limits_{\beta\in J_{r_1,\,n}}\left|\left(\widetilde{\mathbf{A}}^*\widetilde{\mathbf{A}}\right)_{\beta}^{\beta}\right|}\times$$

$$\frac{\sum\limits_{l}\sum\limits_{\alpha\in I_{r_2,\,p}\{l\}}\text{rdet}_l\left(\left(\widetilde{\mathbf{B}}\widetilde{\mathbf{B}}^*\right)_{l.}\left(\widehat{\mathbf{b}}_{s.}\right)\right)_{\alpha}^{\alpha}\cdot m_{lj}^{(\frac{1}{2})}}{\sum\limits_{\alpha\in I_{2,\,p}}\left|\left(\widetilde{\mathbf{B}}\widetilde{\mathbf{B}}^*\right)_{\alpha}^{\alpha}\right|}.$$

Since

$$\widehat{\mathbf{d}}_{.\,s}=\sum_{l=1}^{n}\mathbf{e}_{.\,l}\widehat{d}_{ls},\ \widehat{\mathbf{b}}_{s.}=\sum_{t=1}^{p}\widehat{b}_{st}\mathbf{e}_{t.},\ \sum_{s=1}^{q}\widehat{d}_{ls}\widehat{b}_{st}=\widetilde{d}_{lt},$$

then we have

$$x_{ij}=$$

$$\frac{\sum\limits_{t=1}^{p}\sum\limits_{f=1}^{n}\sum\limits_{k}n_{ik}^{(-\frac{1}{2})}\sum\limits_{\beta\in J_{r_1,\,n}\{k\}}\text{cdet}_k\left(\left(\tilde{\mathbf{A}}^*\tilde{\mathbf{A}}\right)_{.\,k}\left(\mathbf{e}_{.f}\right)\right)_{\beta}^{\beta}\tilde{d}_{ft}\sum\limits_{l}\sum\limits_{\alpha\in I_{r_2,\,p}\{l\}}\text{rdet}_l\left(\left(\tilde{\mathbf{B}}\tilde{\mathbf{B}}^*\right)_{l.}\left(\mathbf{e}_{t.}\right)\right)_{\alpha}^{\alpha}\cdot m_{lj}^{(\frac{1}{2})}}{\sum\limits_{\beta\in J_{r_1,\,n}}\left|\left(\tilde{\mathbf{A}}^*\tilde{\mathbf{A}}\right)_{\beta}^{\beta}\right|\sum\limits_{\alpha\in I_{r_2,p}}\left|\left(\tilde{\mathbf{B}}\tilde{\mathbf{B}}^*\right)_{\alpha}^{\alpha}\right|}. \tag{94}$$

Denote by

$$d_{it}^{\mathbf{A}}:=\sum_{k}n_{ik}^{(-\frac{1}{2})}\sum_{\beta\in J_{r_1,\,n}\{k\}}\text{cdet}_k\left(\left(\widetilde{\mathbf{A}}^*\widetilde{\mathbf{A}}\right)_{.\,k}\left(\widetilde{\mathbf{d}}_{.t}\right)\right)_{\beta}^{\beta}=$$

$$\sum_{f=1}^{n}\sum_{k}n_{ik}^{(-\frac{1}{2})}\sum_{\beta\in J_{r_1,\,n}\{k\}}\text{cdet}_k\left(\left(\widetilde{\mathbf{A}}^*\widetilde{\mathbf{A}}\right)_{.\,k}\left(\mathbf{e}_{.f}\right)\right)_{\beta}^{\beta}\widetilde{d}_{ft}$$

the tth component of the row-vector $\mathbf{d}_{i.}^{\mathbf{A}}=(d_{i1}^{\mathbf{A}},...,d_{ip}^{\mathbf{A}})$ for all $t=1,...,p$. Substituting it in (94), we have

$$x_{ij}=\frac{\sum\limits_{t=1}^{p}d_{it}^{\mathbf{A}}\sum\limits_{l}\sum\limits_{\alpha\in I_{r_2,\,p}\{l\}}\text{rdet}_l\left(\left(\widetilde{\mathbf{B}}\widetilde{\mathbf{B}}^*\right)_{l.}\left(\mathbf{e}_{t.}\right)\right)_{\alpha}^{\alpha}\cdot m_{lj}^{(\frac{1}{2})}}{\sum\limits_{\beta\in J_{r_1,\,n}}\left|\left(\widetilde{\mathbf{A}}^*\widetilde{\mathbf{A}}\right)_{\beta}^{\beta}\right|\sum\limits_{\alpha\in I_{r_2,p}}\left|\left(\widetilde{\mathbf{B}}\widetilde{\mathbf{B}}^*\right)_{\alpha}^{\alpha}\right|}.$$

Since $\sum\limits_{t=1}^{p} d_{it}^{\mathbf{A}} \mathbf{e}_{t.} = \mathbf{d}_{i.}^{\mathbf{A}}$, then it follows (79).

If we denote by

$$\sum_{t=1}^{p} \widetilde{d}_{ft} \sum_{l} \sum_{\alpha \in I_{r_2,\, p}\{l\}} \mathrm{rdet}_l \left(\left(\widetilde{\mathbf{B}}\widetilde{\mathbf{B}}^* \right)_{l.} (\mathbf{e}_{t.}) \right)_{\alpha}^{\alpha} \cdot m_{lj}^{\left(\frac{1}{2}\right)} =$$

$$\sum_{l} \sum_{\alpha \in I_{r_2,\, p}\{l\}} \mathrm{rdet}_l \left(\left(\widetilde{\mathbf{B}}\widetilde{\mathbf{B}}^* \right)_{l.} (\widetilde{\mathbf{d}}_{f.}) \right)_{\alpha}^{\alpha} \cdot m_{lj}^{\left(\frac{1}{2}\right)} =: d_{fj}^{\mathbf{B}}$$

the fth component of the column-vector $\mathbf{d}_{.j}^{\mathbf{B}} = (d_{1j}^{\mathbf{B}}, ..., d_{jn}^{\mathbf{B}})^T$ for all $f = 1, ..., n$ and substitute it in (94), then

$$x_{ij} = \frac{\sum\limits_{f=1}^{n} \sum\limits_{k} n_{ik}^{\left(-\frac{1}{2}\right)} \sum\limits_{\beta \in J_{r_1,\, n}\{k\}} \mathrm{cdet}_k \left(\left(\widetilde{\mathbf{A}}^* \widetilde{\mathbf{A}} \right)_{.k} (\mathbf{e}_{.f}) \right)_{\beta}^{\beta} d_{fj}^{\mathbf{B}}}{\sum\limits_{\beta \in J_{r_1,\, n}} \left| \left(\mathbf{A}^{\sharp}\mathbf{A} \right)_{\beta}^{\beta} \right| \sum\limits_{\alpha \in I_{r_2, p}} \left| \left(\mathbf{B}\mathbf{B}^{\sharp} \right)_{\alpha}^{\alpha} \right|}.$$

Since $\sum\limits_{f=1}^{n} \mathbf{e}_{.f} d_{fj}^{\mathbf{B}} = \mathbf{d}_{.j}^{\mathbf{B}}$, then it follows (78).

(ii) If $\operatorname{rank} \mathbf{A} = n$ and $\operatorname{rank} \mathbf{B} = p$, then by Theorem 4.2 the weighted Moore-Penrose inverses $\mathbf{A}_{M,N}^{\dagger} = \left(a_{ij}^{\ddagger} \right) \in \mathbb{H}^{n \times m}$ and $\mathbf{B}_{P,Q}^{\dagger} = \left(b_{ij}^{\ddagger} \right) \in \mathbb{H}^{q \times p}$ possess the following determinantal representations, respectively,

$$a_{ij}^{\ddagger} = \frac{\mathrm{cdet}_i (\mathbf{A}^*\mathbf{M}\mathbf{A})_{.i} (\widehat{\mathbf{a}}_{.j})}{\det(\mathbf{A}^*\mathbf{M}\mathbf{A})}, \tag{95}$$

$$b_{ij}^{\ddagger} = \frac{\mathrm{rdet}_j (\mathbf{B}\mathbf{Q}^{-1}\mathbf{B}^*)_{j.} (\widehat{\mathbf{b}}_{i.})}{\det(\mathbf{B}\mathbf{Q}^{-1}\mathbf{B}^*)}. \tag{96}$$

where $\widehat{\mathbf{a}}_{.j}$ is the jth column of $\mathbf{A}^*\mathbf{M}$ for all $j = 1, \ldots, m$, and $\widehat{\mathbf{b}}_{i.}$ is the ith row of $\mathbf{Q}^{-1}\mathbf{B}^*$ for all $i = 1, \ldots, n$.

By their substituting in (92), we obtain

$$x_{ij} = \frac{\sum\limits_{t=1}^{p} \sum\limits_{f=1}^{n} \mathrm{cdet}_i (\mathbf{A}^*\mathbf{M}\mathbf{A})_{.i} (\mathbf{e}_{.f}) \, \widetilde{d}_{ft} \mathrm{rdet}_j (\mathbf{B}\mathbf{Q}^{-1}\mathbf{B}^*)_{j.} (\mathbf{e}_{t.})}{\det\, (\mathbf{A}^*\mathbf{M}\mathbf{A}) \det(\mathbf{B}\mathbf{Q}^{-1}\mathbf{B}^*)},$$

where $\widetilde{d}_{ft}$ is the (ft)th entry of $\widetilde{\mathbf{D}} := \mathbf{A}^*\mathbf{M}\mathbf{D}\mathbf{Q}^{-1}\mathbf{B}^*$ in this case. Denote by

$$d_{it}^{\mathbf{A}} := \mathrm{cdet}_i (\mathbf{A}^*\mathbf{M}\mathbf{A})_{.i} (\widetilde{\mathbf{d}}_{.t})$$

the tth component of the row-vector $\mathbf{d}_{i.}^{\mathbf{A}} = (d_{i1}^{\mathbf{A}}, ..., d_{ip}^{\mathbf{A}})$ for all $t = 1, ..., p$. Substituting it in (94), it follows (82).

Similarly, we can obtain (83).

(iii) If $\mathbf{A} \in \mathbb{H}_{r_1}^{m\times n}$, $\mathbf{B} \in \mathbb{H}_{r_2}^{p\times q}$ and $r_1 = n$, $r_2 < p$, then, for the weighted Moore-Penrose inverses $\mathbf{A}_{M,N}^{\dagger}$ and $\mathbf{B}_{P,Q}^{\dagger}$, the determinantal representations (95) and (90) are more applicable to use, respectively. By their substituting in (92) and pondering ahead as in the previous case, we finally obtain (86) and (87) as well.

(iv) In this case for $\mathbf{A}_{M,N}^{\dagger}$ and $\mathbf{B}_{P,Q}^{\dagger}$, we use the determinantal representations (90) and (96), respectively. □

Corollary 6.3. *Suppose that* $\mathbf{A} \in \mathbb{H}_{r_1}^{m\times n}$, $\mathbf{D} \in \mathbb{H}^{m\times p}$, $\mathbf{M}$, $\mathbf{N}$ *are Hermitian positive definite matrices of order* m *and* n, *respectively, and* $\mathbf{A}^{\sharp}\mathbf{A}$ *is non-Hermitian. If* $\mathbf{D} \subset \mathcal{R}_r(\mathbf{A}\mathbf{A}^{\sharp})$ *and* $\mathbf{D} \subset \mathcal{R}_l(\mathbf{A}^{\sharp}\mathbf{A})$, *then the unique solution* $\mathbf{X} = \mathbf{A}_{M,N}^{\dagger}\mathbf{D}$ *of the equation* $\mathbf{AX} = \mathbf{D}$ *with the restrictions (75) possess the following determinantal representations.*

(i) If $\operatorname{rank}\mathbf{A} = r_1 < n$, *then*

$$x_{ij} = \frac{\sum\limits_{k} n_{ik}^{(-\frac{1}{2})} \sum\limits_{\beta\in J_{r_1,\,n}\{i\}} \operatorname{cdet}_k\left(\left(\widetilde{\mathbf{A}}^*\widetilde{\mathbf{A}}\right)_{.k}\left(\widetilde{\mathbf{d}}_{.j}\right)\right)_{\beta}^{\beta}}{\sum\limits_{\beta\in J_{r_1,\,n}} \left|\left(\widetilde{\mathbf{A}}^*\widetilde{\mathbf{A}}\right)_{\beta}^{\beta}\right|},$$

where $\widetilde{\mathbf{d}}_{.j}$ *are the* j*th column of* $\widetilde{\mathbf{D}} = \mathbf{N}^{-\frac{1}{2}}\mathbf{A}^*\mathbf{MD}$ *for all* $i = 1, ..., n$, $j = 1, ..., p$.

(ii) If $\operatorname{rank}\mathbf{A} = n$, *then*

$$x_{ij} = \frac{\operatorname{cdet}_i\left(\mathbf{A}^*\mathbf{MA}\right)_{.i}\left(\widetilde{\mathbf{d}}_{.j}\right)}{\det\left(\mathbf{A}^*\mathbf{MA}\right)},$$

where $\widetilde{\mathbf{d}}_{.j}$ *are the* j*th column of* $\widetilde{\mathbf{D}} = \mathbf{A}^*\mathbf{MD}$.

Proof. The proof follows evidently from Theorem 6.2 when $\mathbf{B}$ be removed and unit matrices insert instead $\mathbf{P}$, $\mathbf{Q}$.

Corollary 6.4. *Suppose that* $\mathbf{B} \in \mathbb{H}^{p\times q}_{r_2}$, $\mathbf{D} \in \mathbb{H}^{n\times q}$, $\mathbf{P}$, *and* $\mathbf{Q}$ *are Hermitian positive definite matrices of order* p *and* q*, respectively, and* $\mathbf{B}\mathbf{B}^\sharp$ *is non-Hermitian. If* $\mathbf{D} \subset \mathcal{R}_r(\mathbf{B}^\sharp\mathbf{B})$ *and* $\mathbf{D} \subset \mathcal{R}_l(\mathbf{B}\mathbf{B}^\sharp)$*, then the unique solution* $\mathbf{X} = \mathbf{D}\mathbf{B}^\dagger_{P,Q}$ *of the equation* $\mathbf{X}\mathbf{B} = \mathbf{D}$ *with the restrictions (77) possess the following determinantal representations.*

(i) *If* $\operatorname{rank}\mathbf{B} = r_2 < p$*, then*

$$x_{ij} = \frac{\sum\limits_{\alpha\in I_{r_2,q}\{j\}} \operatorname{rdet}_j\left(\left(\widetilde{\mathbf{B}}\widetilde{\mathbf{B}}^*\right)_{j.}\left(\widetilde{\mathbf{d}}_{i.}\right)\right)^{\alpha}_{\alpha}}{\sum\limits_{\alpha\in I_{r_2,q}} \left|\left(\widetilde{\mathbf{B}}\widetilde{\mathbf{B}}^*\right)^{\alpha}_{\alpha}\right|},$$

where $\widetilde{\mathbf{d}}_{i.}$ *are the* i*th row of* $\widetilde{\mathbf{D}} = \mathbf{D}\mathbf{Q}^{-1}\mathbf{B}^*\mathbf{P}^{\frac{1}{2}}$ *for all* $i = 1,...,n$, $j = 1,...,p$.

(ii) *If* $\operatorname{rank}\mathbf{B} = p$*, then*

$$x_{ij} = \frac{\operatorname{rdet}_j\left(\mathbf{B}\mathbf{Q}^{-1}\mathbf{B}^*\right)_{j.}\left(\widetilde{\mathbf{d}}_{i.}\right)}{\det\left(\mathbf{B}\mathbf{Q}^{-1}\mathbf{B}^*\right)}, \tag{97}$$

where $\widetilde{\mathbf{d}}_{i.}$ *are the* i*th row of* $\widetilde{\mathbf{D}} = \mathbf{D}\mathbf{Q}^{-1}\mathbf{B}^*$.

Proof. The proof follows evidently from Theorem 6.2 when $\mathbf{A}$ be removed and unit matrices insert instead $\mathbf{M}$, $\mathbf{N}$.

6.3. Mixed Cases

In this subsection we consider mixed cases when only one from the pair $\mathbf{A}^\sharp\mathbf{A}$ and $\mathbf{B}\mathbf{B}^\sharp$ is non-Hermitian. We give this theorems without proofs, since their proofs are similar to the proof of Theorems 6.1 and 6.2.

Theorem 6.3. *Let* $\mathbf{A}^\sharp\mathbf{A}$ *be Hermitian and* $\mathbf{B}\mathbf{B}^\sharp$ *be non-Hermitian. Then the solution (54) possess the following determinantal representations.*

(i) *If* $\operatorname{rank}\mathbf{A} = r_1 < n$ *and* $\operatorname{rank}\mathbf{B} = r_2 < p$*, then*

$$x_{ij} = \frac{\sum\limits_{\beta\in J_{r_1,n}\{i\}} \operatorname{cdet}_i\left(\left(\mathbf{A}^\sharp\mathbf{A}\right)_{.i}\left(\mathbf{d}^{\mathbf{B}}_{.j}\right)\right)^{\beta}_{\beta}}{\sum\limits_{\beta\in J_{r_1,n}} \left|\left(\mathbf{A}^\sharp\mathbf{A}\right)^{\beta}_{\beta}\right| \sum\limits_{\alpha\in I_{r_2,p}} \left|\left(\widetilde{\mathbf{B}}\widetilde{\mathbf{B}}^*\right)^{\alpha}_{\alpha}\right|},$$

or

$$x_{ij} = \frac{\sum\limits_{l} \sum\limits_{\alpha \in I_{r_2,\, p}\{l\}} \operatorname{rdet}_l \left(\left(\widetilde{\mathbf{B}}\widetilde{\mathbf{B}}^* \right)_{l.} \left(\mathbf{d}^{\mathbf{A}}_{i.} \right) \right)^{\alpha}_{\alpha} \cdot m_{lj}^{(\frac{1}{2})}}{\sum\limits_{\beta \in J_{r_1,n}} \left| (\mathbf{A}^{\sharp}\mathbf{A})^{\beta}_{\beta} \right| \sum\limits_{\alpha \in I_{r_2,p}} \left| \left(\widetilde{\mathbf{B}}\widetilde{\mathbf{B}}^* \right)^{\alpha}_{\alpha} \right|},$$

where

$$\mathbf{d}^{\mathbf{B}}_{.j} = \left(\sum_{l} \sum_{\alpha \in I_{r_2,\, p}\{l\}} \operatorname{rdet}_l \left(\left(\widetilde{\mathbf{B}}\widetilde{\mathbf{B}}^* \right)_{l.} (\widetilde{\mathbf{d}}_{t.}) \right)^{\alpha}_{\alpha} \cdot m_{lj}^{(\frac{1}{2})} \right) \in \mathbb{H}^{n \times 1} \quad (98)$$

$$\mathbf{d}^{\mathbf{A}}_{i.} = \left(\sum_{\beta \in J_{r_1,n}\{i\}} \operatorname{cdet}_i \left(\left(\mathbf{A}^{\sharp}\mathbf{A} \right)_{.i} \left(\widetilde{\mathbf{d}}_{.f} \right) \right)^{\beta}_{\beta} \right) \in \mathbb{H}^{1 \times p} \quad (99)$$

are the column-vector and the row-vector, respectively. $\tilde{\mathbf{d}}_{t.}$ *and* $\widetilde{\mathbf{d}}_{.f}$ *are the* t*th row and the* f*th column of* $\widetilde{\mathbf{D}} := \mathbf{A}^{\sharp}\mathbf{D}\mathbf{Q}^{-1}\mathbf{B}^*\mathbf{P}^{\frac{1}{2}}$ *for all* $t = 1, ..., n$, $f = 1, ..., p$.

(ii) If $\operatorname{rank} \mathbf{A} = n$ *and* $\operatorname{rank} \mathbf{B} = p$*, then*

$$x_{ij} = \frac{\operatorname{cdet}_i (\mathbf{A}^{\sharp}\mathbf{A})_{.i} \left(\mathbf{d}^{\mathbf{B}}_{.j} \right)}{\det(\mathbf{A}^{\sharp}\mathbf{A}) \cdot \det(\mathbf{B}\mathbf{Q}^{-1}\mathbf{B}^*)},$$

or

$$x_{ij} = \frac{\operatorname{rdet}_j (\mathbf{B}\mathbf{Q}^{-1}\mathbf{B}^*)_{j.} \left(\mathbf{d}^{\mathbf{A}}_{i.} \right)}{\det(\mathbf{A}^{\sharp}\mathbf{A}) \cdot \det(\mathbf{B}\mathbf{Q}^{-1}\mathbf{B}^*)},$$

where

$$\mathbf{d}^{\mathbf{B}}_{.j} := \left(\operatorname{rdet}_j (\mathbf{B}\mathbf{Q}^{-1}\mathbf{B}^*)_{j.} \left(\tilde{\mathbf{d}}_{t.} \right) \right) \in \mathbb{H}^{n \times 1}, \quad (100)$$

$$\mathbf{d}^{\mathbf{A}}_{i.} := \left(\operatorname{cdet}_i (\mathbf{A}^{\sharp}\mathbf{A})_{.i} \left(\tilde{\mathbf{d}}_{.f} \right) \right) \in \mathbb{H}^{1 \times p}, \quad (101)$$

$\tilde{\mathbf{d}}_{t.}$, $\tilde{\mathbf{d}}_{.f}$ *are the* t*th row and* f*th column of* $\tilde{\mathbf{D}} = \mathbf{A}^{\sharp}\mathbf{D}\mathbf{Q}^{-1}\mathbf{B}^*$.

(iii) If $\operatorname{rank} \mathbf{A} = n$ *and* $\operatorname{rank} \mathbf{B} = r_2 < p$*, then*

$$x_{ij} = \frac{\operatorname{cdet}_i \left(\left(\mathbf{A}^{\sharp}\mathbf{A} \right)_{.i} \left(\mathbf{d}^{\mathbf{B}}_{.j} \right) \right)}{\det(\mathbf{A}^{\sharp}\mathbf{A}) \cdot \sum\limits_{\alpha \in I_{r_2,p}} \left| \left(\widetilde{\mathbf{B}}\widetilde{\mathbf{B}}^* \right)^{\alpha}_{\alpha} \right|},$$

or

$$x_{ij} = \frac{\sum\limits_{l}\sum\limits_{\alpha\in I_{r_2,p}\{l\}} \mathrm{rdet}_l\left(\left(\widetilde{\mathbf{B}}\widetilde{\mathbf{B}}^*\right)_{l.}\left(\mathbf{d}_{i.}^{\mathbf{A}}\right)\right)_{\alpha}^{\alpha}\cdot m_{lj}^{(\frac{1}{2})}}{\det(\mathbf{A}^\sharp\mathbf{A})\cdot\sum\limits_{\alpha\in I_{r_2,p}}\left|\left(\widetilde{\mathbf{B}}\widetilde{\mathbf{B}}^*\right)_{\alpha}^{\alpha}\right|},$$

where $\mathbf{d}_{.j}^{\mathbf{B}}$ *is (98) and* $\mathbf{d}_{i.}^{\mathbf{A}}$ *is (101).*

(iv) If $\mathrm{rank}\,\mathbf{A} = r_1 < n$ *and* $\mathrm{rank}\,\mathbf{B} = p$*, then*

$$x_{ij} = \frac{\mathrm{rdet}_j(\mathbf{B}\mathbf{Q}^{-1}\mathbf{B}^*)_{j.}\left(\mathbf{d}_{i.}^{\mathbf{A}}\right)}{\sum\limits_{\beta\in J_{r_1,n}}\left|(\mathbf{A}^\sharp\mathbf{A})_{\beta}^{\beta}\right|\cdot\det(\mathbf{B}\mathbf{Q}^{-1}\mathbf{B}^*)},$$

or

$$x_{ij} = \frac{\sum\limits_{\beta\in J_{r_1,n}\{i\}} \mathrm{cdet}_i\left((\mathbf{A}^\sharp\mathbf{A})_{.i}\left(\mathbf{d}_{.j}^{\mathbf{B}}\right)\right)_{\beta}^{\beta}}{\sum\limits_{\beta\in J_{r_1,n}}\left|(\mathbf{A}^\sharp\mathbf{A})_{\beta}^{\beta}\right|\cdot\det(\mathbf{B}\mathbf{Q}^{-1}\mathbf{B}^*)},$$

where $\mathbf{d}_{.j}^{\mathbf{B}}$ *is (100) and* $\mathbf{d}_{i.}^{\mathbf{A}}$ *is (99).*

Theorem 6.4. *Let* $\mathbf{A}^\sharp\mathbf{A}$ *be non-Hermitian, and* $\mathbf{B}\mathbf{B}^\sharp$ *be Hermitian. Denote* $\widetilde{\mathbf{D}} := \widetilde{\mathbf{A}}^*\mathbf{D}\mathbf{B}^\sharp$*. Then the solution (54) possess the following determinantal representations.*

(i) If $\mathrm{rank}\,\mathbf{A} = r_1 < n$ *and* $\mathrm{rank}\,\mathbf{B} = r_2 < p$*, then*

$$x_{ij} = \frac{\sum\limits_{k} n_{ik}^{(-\frac{1}{2})}\sum\limits_{\beta\in J_{r_1,n}\{k\}} \mathrm{cdet}_k\left(\left(\widetilde{\mathbf{A}}^*\widetilde{\mathbf{A}}\right)_{.k}\left(\mathbf{d}_{.j}^{\mathbf{B}}\right)\right)_{\beta}^{\beta}}{\sum\limits_{\beta\in J_{r_1,n}}\left|\left(\widetilde{\mathbf{A}}^*\widetilde{\mathbf{A}}\right)_{\beta}^{\beta}\right|\sum\limits_{\alpha\in I_{r_2,p}}\left|(\mathbf{B}\mathbf{B}^\sharp)_{\alpha}^{\alpha}\right|},$$

or

$$x_{ij} = \frac{\sum\limits_{\alpha\in I_{r_2,p}\{j\}} \mathrm{rdet}_j\left((\mathbf{B}\mathbf{B}^\sharp)_{j.}\left(\mathbf{d}_{i.}^{\mathbf{A}}\right)\right)_{\alpha}^{\alpha}}{\sum\limits_{\beta\in J_{r_1,n}}\left|\left(\widetilde{\mathbf{A}}^*\widetilde{\mathbf{A}}\right)_{\beta}^{\beta}\right|\sum\limits_{\alpha\in I_{r_2,p}}\left|(\mathbf{B}\mathbf{B}^\sharp)_{\alpha}^{\alpha}\right|},$$

where

$$\mathbf{d}^{\mathbf{B}}_{.j} = \left(\sum_{\alpha \in I_{r_2,p}\{j\}} \mathrm{rdet}_j \left(\left(\mathbf{B}\mathbf{B}^{\sharp} \right)_{j.} \left(\tilde{\mathbf{d}}_{t.} \right) \right) {}^{\alpha}_{\alpha} \right) \in \mathbb{H}^{n\times 1} \tag{102}$$

$$\mathbf{d}^{\mathbf{A}}_{i.} = \left(\sum_{k} n^{(-\frac{1}{2})}_{ik} \sum_{\beta \in J_{r_1,\,n}\{k\}} \mathrm{cdet}_k \left(\left(\widetilde{\mathbf{A}}^{*} \widetilde{\mathbf{A}} \right)_{.\,k} \left(\tilde{\mathbf{d}}_{.f} \right) \right)^{\beta}_{\beta} \right) \in \mathbb{H}^{1\times p} \tag{103}$$

are the column-vector and the row-vector, respectively. $\tilde{\mathbf{d}}_{t.}$ *and* $\tilde{\mathbf{d}}_{.f}$ *are the tth row and the fth column of* $\widetilde{\mathbf{D}} := \mathbf{N}^{-\frac{1}{2}}\mathbf{A}^{*}\mathbf{M}\mathbf{D}\mathbf{B}^{\sharp}$ *for all* $t = 1, ..., n$, $f = 1, ..., p$.

(ii) If $\mathrm{rank}\,\mathbf{A} = n$ *and* $\mathrm{rank}\,\mathbf{B} = p$*, then*

$$x_{ij} = \frac{\mathrm{cdet}_i (\mathbf{A}^{*}\mathbf{M}\mathbf{A})_{.i} \left(\mathbf{d}^{\mathbf{B}}_{.j} \right)}{\det(\mathbf{A}^{*}\mathbf{M}\mathbf{A}) \cdot \det\left(\mathbf{B}\mathbf{B}^{\sharp}\right)},$$

or

$$x_{ij} = \frac{\mathrm{rdet}_j \left(\mathbf{B}\mathbf{B}^{\sharp}\right)_{j.} \left(\mathbf{d}^{\mathbf{A}}_{i.} \right)}{\det(\mathbf{A}^{*}\mathbf{M}\mathbf{A}) \cdot \det\left(\mathbf{B}\mathbf{B}^{\sharp}\right)},$$

where

$$\mathbf{d}^{\mathbf{B}}_{.j} := \left(\mathrm{rdet}_j \left(\mathbf{B}\mathbf{B}^{\sharp} \right)_{j.} \left(\tilde{\mathbf{d}}_{t.} \right) \right) \in \mathbb{H}^{n\times 1}, \tag{104}$$

$$\mathbf{d}^{\mathbf{A}}_{i.} := \left(\mathrm{cdet}_i (\mathbf{A}^{*}\mathbf{M}\mathbf{A})_{.i} \left(\tilde{\mathbf{d}}_{.f} \right) \right) \in \mathbb{H}^{1\times p}, \tag{105}$$

$\tilde{\mathbf{d}}_{t.}$, $\tilde{\mathbf{d}}_{.f}$ *are the tth row and fth column of* $\tilde{\mathbf{D}} = \mathbf{A}^{*}\mathbf{M}\mathbf{D}\mathbf{B}^{\sharp}$.

(iii) If $\mathrm{rank}\,\mathbf{A} = n$ *and* $\mathrm{rank}\,\mathbf{B} = r_2 < p$*, then*

$$x_{ij} = \frac{\mathrm{cdet}_i \left((\mathbf{A}^{*}\mathbf{M}\mathbf{A})_{.i} \left(\mathbf{d}^{\mathbf{B}}_{.j} \right) \right)}{\det(\mathbf{A}^{*}\mathbf{M}\mathbf{A}) \cdot \sum\limits_{\alpha \in I_{r_2,p}} \left| (\mathbf{B}\mathbf{B}^{\sharp})^{\alpha}_{\alpha} \right|},$$

or

$$x_{ij} = \frac{\sum\limits_{\alpha \in I_{r_2,p}\{j\}} \mathrm{rdet}_j \left(\left(\mathbf{B}\mathbf{B}^{\sharp} \right)_{j.} \left(\mathbf{d}^{\mathbf{A}}_{i.} \right) \right) {}^{\alpha}_{\alpha}}{\det(\mathbf{A}^{*}\mathbf{M}\mathbf{A}) \cdot \sum\limits_{\alpha \in I_{r_2,p}} \left| (\mathbf{B}\mathbf{B}^{\sharp})^{\alpha}_{\alpha} \right|},$$

where $\mathbf{d}^{\mathbf{B}}_{.j}$ *is (102) and* $\mathbf{d}^{\mathbf{A}}_{i.}$ *is (105).*

(iv) If $\operatorname{rank}\mathbf{A} = r_1 < n$ *and* $\operatorname{rank}\mathbf{B} = p$*, then*

$$x_{ij} = \frac{\operatorname{rdet}_j\left(\mathbf{B}\mathbf{B}^{\sharp}\right)_{j.}\left(\mathbf{d}^{\mathbf{A}}_{i\,.}\right)}{\sum\limits_{\beta\in J_{r_1,n}}\left|\left(\widetilde{\mathbf{A}}^*\widetilde{\mathbf{A}}\right)^{\beta}_{\beta}\right|\cdot\det\left(\mathbf{B}\mathbf{B}^{\sharp}\right)},$$

or

$$x_{ij} = \frac{\sum\limits_{\beta\in J_{r_1,\,n}\{i\}}\operatorname{cdet}_i\left(\left(\mathbf{A}^{\sharp}\mathbf{A}\right)_{.\,i}\left(\mathbf{d}^{\,\mathbf{B}}_{.j}\right)\right)^{\beta}_{\beta}}{\sum\limits_{\beta\in J_{r_1,n}}\left|\left(\widetilde{\mathbf{A}}^*\widetilde{\mathbf{A}}\right)^{\beta}_{\beta}\right|\cdot\det(\mathbf{B}\mathbf{Q}^{-1}\mathbf{B}^*)},$$

where $\mathbf{d}^{\mathbf{B}}_{.j}$ *is (104) and* $\mathbf{d}^{\mathbf{A}}_{i\,.}$ *is (103).*

7. Examples

In this section, we give examples to illustrate our results.

1. Let us consider the matrices

$$\mathbf{A} = \begin{pmatrix} 1 & i & j \\ -k & i & 1 \\ k & j & -i \\ j & -1 & i \end{pmatrix}, \tag{106}$$

$$\mathbf{N}^{-1} = \begin{pmatrix} 23 & 16-2i-2j+10k & -16+10i-2j-2k \\ 16+2i+2j-10k & 29 & -19-i-13j-k \\ -16-10i+2j+2k & -19+i+13j+k & 29 \end{pmatrix},$$

$$\mathbf{M} = \begin{pmatrix} 2 & k & i & 0 \\ -k & 2 & 0 & j \\ -i & 0 & 2 & k \\ 0 & -j & -k & 2 \end{pmatrix}. \tag{107}$$

By direct calculation we get that leading principal minors of $\mathbf{M}$ and $\mathbf{N}^{-1}$ are all positive. So, $\mathbf{M}$ and $\mathbf{N}^{-1}$ are positive definite matrices. Similarly, by direct calculation of leading principal minors of $\mathbf{A}^*\mathbf{A}$, we obtain $\operatorname{rank}\mathbf{A}^*\mathbf{A} = \operatorname{rank}\mathbf{A} = 2$. Further,

$$\mathbf{A}^{\sharp} = \mathbf{M}\mathbf{A}^*\mathbf{N}^{-1} =$$

$$\begin{pmatrix} 51-12i+25j-24k & -43-18i+39k & -18+26i-30j-38k & 19-i-50j-42k \\ -32i+17j-37k & -24-50i+26j+24k & -5-24i-56j+k & -38-25i-18j-67k \\ 5-6i-50j+11k & 44+23i-12j+7k & 30+38i+5j+37k & 18-44i+6j+54k \end{pmatrix}.$$

Since,

$$\mathbf{A}^{\sharp}\mathbf{A} =$$

$$\begin{pmatrix} 178 & 41+47i+47j+43k & -41+43i+47j+47k \\ 41-47i-47j-43k & 176 & -40-46i-42j-46k \\ -41-43i-47j-47k & -40+46i+42j+46k & 176 \end{pmatrix}$$

are Hermitian, then we shall be obtain $\mathbf{A}^{\dagger}_{M,N} = \left(\tilde{a}^{\dagger}_{ij}\right) \in \mathbb{H}^{3\times 4}$ due to Theorem 4.1 by Eq. (33).

We have, $\sum\limits_{\beta\in J_{2,\,3}} \left|\left(\mathbf{A}^{\sharp}\mathbf{A}\right){}^{\beta}_{\beta}\right| = 23380 + 23380 + 23380 = 70140$, and

$$\sum_{\beta\in J_{2,3}\{1\}} \operatorname{cdet}_1\left(\left(\mathbf{A}^{\sharp}\mathbf{A}\right)_{.1}\left(\mathbf{a}^{\sharp}_{.1}\right)\right){}^{\beta}_{\beta} = 6680 + 1670i + 3340j - 5010k +$$

$$6680 - 5010i + 3340j - 1640k = 13360 - 3340i + 6680j - 6680k.$$

Then,

$$\tilde{a}^{\dagger}_{11} = \frac{8-2i+4j-4k}{42}.$$

Similarly, we obtain

$$\tilde{a}^{\dagger}_{12} = \tfrac{-7-3i+6k}{42},\quad \tilde{a}^{\dagger}_{13} = \tfrac{-3+4i-5j-6k}{42},\quad \tilde{a}^{\dagger}_{14} = \tfrac{3-8j-7k}{42},$$

$$\tilde{a}^{\dagger}_{21} = \tfrac{-5i+3j-6k}{42},\quad \tilde{a}^{\dagger}_{22} = \tfrac{-4-8i+2j+2k}{42},\quad \tilde{a}^{\dagger}_{23} = \tfrac{-1-4i-9j}{42}, \tilde{a}^{\dagger}_{24} = \tfrac{-6-4i-3j-11k}{42},$$

$$\tilde{a}^{\dagger}_{31} = \tfrac{-1-i-8j+2k}{42},\quad \tilde{a}^{\dagger}_{32} = \tfrac{7+4i-2j+k}{42},\quad \tilde{a}^{\dagger}_{33} = \tfrac{5+6i+j+6k}{42}, \tilde{a}^{\dagger}_{34} = \tfrac{3-7i+j+9k}{42}.$$

Finally, we obtain

$$\mathbf{A}^{\dagger}_{M,N} = \frac{1}{42}\begin{pmatrix} 8-2i+4j-4k & -7-3i+6k & -3+4i-5j-6k & 3-8j-7k \\ -5i+3j-6k & -4-8i+2j+2k & -1-4i-9j & -6-4i-3j-11k \\ -1-i-8j+2k & 7+4i-2j+k & 5+6i+j+6k & 3-7i+j+9k \end{pmatrix}. \tag{108}$$

2. Consider the right system of linear equations,

$$\mathbf{A}\mathbf{x} = \mathbf{b}, \tag{109}$$

where the coefficient matrix $\mathbf{A}$ is (106) and the column $\mathbf{b} = (1\ 0\ i\ k)^T$. Using (108), by the matrix method we have for the weighted Moore-Penrose solution $\tilde{\mathbf{x}} = \mathbf{A}^{\dagger}_{M,N}\mathbf{b}$ of (109) with weights $\mathbf{M}$ and $\mathbf{N}$ from (107),

$$\tilde{x}_1 = \frac{11 - 13i - 2j + 4k}{42}, \tilde{x}_2 = \frac{15 - 9i + 7j - 3k}{42}, \tilde{x}_3 = \frac{-16 + 5i + 5j + 4k}{42}. \tag{110}$$

Now, we shall find the weighted Moore-Penrose solution of (109) by Cramer's rule (48). Since

$$\mathbf{f} = \mathbf{A}^{\sharp}\mathbf{b} = \begin{pmatrix} 67 - 80i - 12j + 25k \\ 91 - 55i + 43j - 19k \\ -97 + 30i + 31j + 24k \end{pmatrix},$$

then we have

$$\tilde{x}_1 = \frac{\sum\limits_{\beta \in J_{2,\,3}\{i\}} \operatorname{cdet}_i \left(\left(\mathbf{A}^{\sharp}\mathbf{A} \right)_{.\,i} (\mathbf{f}) \right)^{\beta}_{\beta}}{\sum\limits_{\beta \in J_{r,\,n}} \left| \left(\mathbf{A}^{\sharp}\mathbf{A} \right)^{\beta}_{\beta} \right|} = \frac{18370 - 21710i - 3340j + 6680k}{70140} = \frac{11 - 13i - 2j + 4k}{42},$$

$$\tilde{x}_2 = \frac{25050 - 15030i + 11690j - 5010k}{70140} = \frac{15 - 9i + 7j - 3k}{42},$$
$$\tilde{x}_3 = \frac{-26720 + 8350i + 8350j + 6680k}{70140} = \frac{-16 + 5i + 5j + 4k}{42}.$$

As we expected, the weighted Moore-Penrose solutions by Cramer's rule and be the matrix method coincide.

3. Let us consider the restricted matrix equation

$$\mathbf{XB} = \mathbf{D}, \ \ \mathcal{N}_r(\mathbf{X}) \supset \mathbf{P}^{-1}\mathcal{N}_r(\mathbf{B}^*), \mathcal{N}_l(\mathbf{X}) \supset \mathcal{N}_l(\mathbf{B}^*)\mathbf{Q}, \tag{111}$$

where

$$\mathbf{B} = \begin{pmatrix} k & -j & j \\ 0 & 1 & i \end{pmatrix}, \mathbf{Q} = \begin{pmatrix} 2 & i & k \\ -i & 2 & -j \\ -k & j & 3 \end{pmatrix}, \mathbf{P} = \begin{pmatrix} 1 & j \\ -j & 2 \end{pmatrix}.$$

The inverse $\mathbf{Q}^{-1}$ can be obtain due to Theorem 2.4, then

$$\mathbf{Q}^{-1} = \frac{1}{3}\begin{pmatrix} 5 & -4i & -3k \\ 4i & 5 & 3j \\ 3k & -3j & 3 \end{pmatrix}.$$

Since

$$\mathbf{B}\mathbf{Q}^{-1}\mathbf{B}^*\mathbf{P} = \begin{pmatrix} 4+i-j+2k & 2-4i+j+2k \\ 1+2i+\frac{1}{3}j-k & \frac{7}{3}+i+j+2k \end{pmatrix}$$

is not Hermitian, and

$$\det\left(\mathbf{B}\mathbf{Q}^{-1}\mathbf{B}\right) = \det\begin{pmatrix} 7 & 1-2i-3j+k \\ 1+2i+3j-k & \frac{8}{3} \end{pmatrix} = 11,$$

then we shall find the solution of (111) by (97). Therefore,

$$x_{11} = \frac{\operatorname{rdet}_1\left(\mathbf{B}\mathbf{Q}^{-1}\mathbf{B}^*\right)_{1.}\left(\check{\mathbf{d}}_{i.}\right)}{\det\left(\mathbf{B}\mathbf{Q}^{-1}\mathbf{B}^*\right)} = \frac{1}{11}\operatorname{rdet}_1\begin{pmatrix} 1+i+3j+2k & \frac{4}{3}+i-j+k \\ 1+2i+3j-k & \frac{8}{3} \end{pmatrix} = \frac{1}{33}(-2+3i+6j+2k),$$

because $\widehat{\mathbf{d}}_{1.}$ is the first row of

$$\widehat{\mathbf{D}} = \mathbf{D}\mathbf{Q}^{-1}\mathbf{B}^* = \begin{pmatrix} 1+i+3j+2k & \frac{4}{3}+i-j+k \\ 1+3i+3j-k & \frac{8}{3}-\frac{1}{3}k \\ 1-i+2j-3k & 1-i+j+k \end{pmatrix}.$$

Similarly, we obtain

$$x_{21} = \frac{1}{11}\operatorname{rdet}_1\begin{pmatrix} 1+3i+3j-k & \frac{8}{3}-\frac{1}{3}k \\ 1+2i+3j-k & \frac{8}{3} \end{pmatrix} = \frac{1}{33}(1+5i+2j+k),$$

and

$$\begin{aligned} x_{31} &= \frac{1}{33}(11-17i+13j-3k), & x_{12} &= \frac{1}{11}(-2-3i-6j+3k), \\ x_{22} &= \frac{1}{33}(5-3i+3j+2k), & x_{32} &= \frac{1}{11}(-1+3i+j-12k). \end{aligned}$$

Note that we used Maple with the package CLIFFORD in the calculations.

Conclusion

In this chapter, we derive determinantal representations of the weighted Moore-Penrose by WSVD within the framework of the theory of noncommutative column-row determinants. By using the obtained analogs of the adjoint matrix, we get the Cramer rules for solutions of restricted left and right systems of quaternion linear equations. We give determinantal representations for solutions of quaternion restricted matrix equation $\mathbf{AXB} = \mathbf{D}$ in all cases with respect to weighted matrices.

References

[1] Penrose, R. A. Generalized inverse for matrices, *Proc. Camb. Philos. Soc.* 1955, 51, 406–413.

[2] Prasad, K.M., Bapat, R.B. A note of the Khatri inverse, *Sankhya: Indian J. Stat.* 1992, 54, 291-295.

[3] Wei, Y., Wu, H. The representation and approximation for the weighted Moore-Penrose inverse, *Appl. Math. Comput.* 2001, 121, 17-28.

[4] Sergienko, I.V., Galba, E.F., Deineka, V.S. Limiting representations of weighted pseudoinverse matrices with positive definite weights. Problem regularization, *Cybernetics and Systems Analysis* 2003, 39(6), 816-830.

[5] Stanimirovic', P., Stankovic', M. Determinantal representation of weighted Moore-Penrose inverse, *Mat. Vesnik* 1994, 46, 41-50.

[6] Liu, X., Zhu, G., Zhou, G., Yu, Y. An analog of the adjugate matrix for the outer inverse $\mathbf{A}^{(2)}_{T,S}$, *Math. Problem. in Eng.* 2012, Article ID 591256, 14 pages.

[7] Kyrchei, I.I. Analogs of the adjoint matrix for generalized inverses and corresponding Cramer rules, *Linear and Multilinear Algebra* 2008, 56(4), 453-469.

[8] Liu, X., Yu, Y., Wang, H. Determinantal representation of weighted generalized inverses, *Appl. Math. Comput.* 2011, 218(7), 3110-3121.

[9] Zhang, F. Quaternions and matrices of quaternions, *Linear Algebra Appl.* 1997, 251, 21-57.

[10] Kyrchei, I.I. Determinantal representation of the Moore-Penrose inverse matrix over the quaternion skew field, *J. Math. Sci.* 2012, 180(1), 23-33.

[11] Van Loan, C.F. Generalizing the singular value decomposition, *SIAM J. Numer. Anal.* 1976, 13, 76–83.

[12] Galba, E. F. Weighted singular decomposition and weighted pseudoinversion of matrices, *Ukr. Math. J.* 1996, 48(10), 1618-1622.

[13] Kyrchei, I.I. Cramer's rule for generalized inverse solutions, In *Advances in Linear Algebra Research*, Kyrchei, I.I., (Ed.) Nova Sci. Publ., New York, 2015, pp. 79-132.

[14] Kyrchei, I.I. The theory of the column and row determinants in a quaternion linear algebra, In *Advances in Mathematics Research*, Baswell, A.R, (Ed.) Nova Sci. Publ., New York, 2012, Vol. 15, pp. 301-359.

[15] Kyrchei, I.I. Determinantal representations of the Moore-Penrose inverse over the quaternion skew field and corresponding Cramer's rules, *Linear Multilinear Algebra* 2011, 59(4), 413-431.

[16] Song, G., Wang, Q., Chang, H. Cramer rule for the unique solution of restricted matrix equations over the quaternion skew field, *Comput Math. Appl.* 2011, 61, 1576-1589.

[17] Song, G.J., Wang, Q.W. Condensed Cramer rule for some restricted quaternion linear equations, *Appl. Math. Comput.* 2011, 218, 3110-3121.

[18] Kyrchei, I.I. Weighted singular value decomposition and determinantal representations of the quaternion weighted Moore-Penrose inverse, *Appl. Math. Comput.* 2017, 309, 1-16.

[19] Kyrchei, I.I. Explicit representation formulas for the minimum norm least squares solutions of some quaternion matrix equations, *Linear Algebra Appl.* 2013, 438(1), 136–152.

[20] Kyrchei, I.I. Determinantal representations of the Drazin inverse over the quaternion skew field with applications to some matrix equations, *Appl. Math. Comput.* 2014, 238, 193–207.

[21] Kyrchei, I.I. Determinantal representations of the W-weighted Drazin inverse over the quaternion skew field, *Appl. Math. Comput.* 2015, 264, 453–465.

[22] Kyrchei, I.I. Explicit determinantal representation formulas of W-weighted Drazin inverse solutions of some matrix equations over the quaternion skew field, *Math. Problem. in Eng.* 2016, Article ID 8673809, 13 pages.

[23] Kyrchei, I.I. Determinantal representations of the Drazin and W-weighted Drazin inverses over the quaternion skew field with applications, In *Quaternions: Theory and Applications*, Griffin, S. (Ed.) New York: Nova Sci. Publ. 2017, pp.201-275.

[24] Kleyn A., Kyrchei, I. Relation of row-column determinants with quasideterminants of matrices over a quaternion algebra, In *Advances in Linear Algebra Research*, Kyrchei, I.I., (Ed.) Nova Sci. Publ., New York, 2015, pp. 299-324.

[25] Song, G.J., Dong, C.Z. New results on condensed Cramers rule for the general solution to some restricted quaternion matrix equations, *J. Appl. Math. Comput.* 2017, 53, 321–341.

[26] Song, G.J. Bott-Duffin inverse over the quaternion skew field with applications, *J. Appl. Math. Comput.* 2013, 41, 377-392.

[27] Song, G.J. Characterization of the W-weighted Drazin inverse over the quaternion skew field with applications, *Electron. J. Linear Algebra* 2013, 26, 1–14.

[28] Song, G.J. Determinantal representation of the generalized inverses over the quaternion skew field with applications, *Appl. Math. Comput.* 2012, 219, 656–667.

[29] Huang, L., So, W. On left eigenvalues of a quaternionic matrix, *Linear Algebra Appl.* 2001, 323, 105-116.

[30] So, W. Quaternionic left eigenvalue problem, *Southeast Asian Bulletin of Mathematics* 2005, 29, 555-565.

[31] Wood, R. M. W. Quaternionic eigenvalues, *Bull. Lond. Math. Soc.* 1985, 17, 137-138.

[32] Brenner, J.L. Matrices of quaternions , *Pac. J. Math.* 1951, 1, 329-335.

[33] Macías-Virgós, E., Pereira-Sáez, M.J. A topological approach to left eigenvalues of quaternionic matrices, *Linear Multilinear Algebra* 2014, 62(2), 139–158.

[34] Baker, A. Right eigenvalues for quaternionic matrices: a topological approach, *Linear Algebra Appl.* 1999, 286, 303-309.

[35] Dray, T., Manogue, C. A. The octonionic eigenvalue problem, Advances in *Applied Clifford Algebras* 1998, 8(2), 341-364.

[36] Zhang, F. Quaternions and matrices of quaternions, *Linear Algebra Appl.* 1997, 251, 21-57.

[37] Farenick, D.R., Pidkowich, B.A.F. The spectral theorem in quaternions, *Linear Algebra Appl.* 2003, 371, 75-102.

[38] Farid, F. O., Wang, Q.W., Zhang, F. On the eigenvalues of quaternion matrices, *Linear Multilinear Algebra* 2011, 59(4), 451–473.

[39] Horn, R. A., Johnson, C. R. *Matrix Analysis*. Cambridge etc., Cambridge University Press, 1985.

[40] Ben-Israel, A., Grenville, T.N.E. *Generalized Inverses: Theory and Applications.* Springer- Verlag, Berlin, 2002.

In: Advances in Mathematics Research
Editor: Albert R. Baswell

ISBN: 978-1-53612-512-2

Chapter 3

SADDLE POINTS CRITERIA FOR A NEW CLASS OF NONCONVEX NONSMOOTH DISCRETE MINIMAX FRACTIONAL PROGRAMMING PROBLEMS

Tadeusz Antczak*
Faculty of Mathematics and Computer Science
University of Łódź, Łódź, Poland

Abstract

In this paper, we present several sets of saddle point criteria for a new class of nonconvex nonsmooth discrete minimax fractional programming problems in which the involving functions are (Ψ, Φ, ρ)-univex and/or (Ψ, Φ, ρ)-pseudounivex. The results extend and generalize the corresponding results established earlier in the literature for such nonsmooth optimization problems.

Keywords: minmax fractional programming problem, saddle point, Lagrange function, nonsmooth (Ψ, Φ, ρ)-univex function, nonsmooth (Ψ, Φ, ρ)-pseudounivex function

AMS Classification: 90C47, 90C32, 90C26, 90C30, 90C46

*Corresponding Author Email: antczak@math.uni.lodz.pl.

1. Introduction

In this paper, we will use seven seemingly different but essentially equivalent Lagrangian-type functions to state and prove four sets of saddle-point-type necessary and sufficient optimality conditions under appropriate (Ψ, Φ, ρ)-univexity and/or (Ψ, Φ, ρ)-pseudounivexity assumptions for the following discrete minimax fractional programming problem

$$\begin{aligned} &\min_{x \in X} \max\nolimits_{1 \le i \le p} \frac{f_i(x)}{q_i(x)} \\ \text{subject to } & g_j(x) \le 0, \ j \in J = \{1, ..., m\}, \\ & h_k(x) = 0, \ k \in K = \{1, ..., r\}, \\ & x \in X, \end{aligned} \qquad \text{(P)}$$

where X is a real Banach space, for each $i \in I \equiv \{1, 2, \ldots, p\}$, f_i and q_i are real-valued functions defined on X, $q_i(x) > 0$, $i \in I$, for all x satisfying the constraints of (P), for each $j \in J$, g_j is a real-valued function defined on X and, for each $k \in K$, h_k is a real-valued function defined on X. By D, we denote the set of all feasible solutions in the considered minimax fractional programming problem (P), that is,

$$D = \{x \in X : q_i(x) > 0, i \in I, g_j(x) \le 0, \ j \in J, h_k(x) = 0, \ k \in K\}.$$

Minimax fractional programming problems have been became a subject of wide interest recently since they provide a universal apparatus for a wide class of models in economics, decision making, game theory, engineering design, health care, transportation, and others. In the last few years, many concepts of generalized convexity have been introduced and applied to different classes of nonconvex minimax fractional programming problems for proving a variety optimality and duality results for such optimization problems (see, for example, [1], [2], [3], [4], [5], [6], [7], [10], [17], [18], [19], [20], [21], [22], [23], [24], [25], [26], [28], [29], [35], [38], [41], [43], and others).

In the context of nonlinear programming theory, the characterization of a constrained optimum as a saddle point of the Lagrangian function is known in the literature to be heavily dependent upon convexity properties of the underlying optimization problem (see, for example, [11], [33]). In the optimization literature, however, there are only very few works available dealing with saddle point criteria for nonconvex minimax fractional programming problems (see, for example, [8], [30], [32], [34], [39]). In [39], Zalmai derived saddle point criteria for a class of nondifferentiable minimax programming problems with generalized convex functions. Xu [36] presented the so-called GK-saddle-point type optimality criteria for generalized minimax fractional programming problem using the introduced functions somewhat like the Lagrangian functions. Chandra and Kumar [14] considered different Lagrangian functions and, under convexity, established saddle point type optimality criteria for generalized fractional programming problem. In [32], Luo defined the incomplete Lagrange function and using it he proved saddle point optimality criteria for a class of nondifferentiable generalized fractional programming. Yang et al. [37] established two theorems of alternative with generalized subconvexlikeness and they then used them to establish a saddle-point type optimality conditions for the considered generalized fractional programming problem. Reddy et al. [34] proved saddle-point optimality criteria and duality results for a class of nonconvex minimax multiobjective fractional programming problem. Zalmai [42] gave several sets of saddle-point-type optimality conditions and proved weak and strong duality theorems between a discrete minmax fractional subset programming problem and its two Lagrangian-type dual problems under appropriate $(b, \varphi, \rho, \theta)$-convexity hypotheses. Recently, Antczak [8] considered a class of nonconvex semi-infinite minimax fractional programming problem with both inequality and equality constraints. He gave characterizations of an optimal solution by a saddle point of the scalar Lagrange function and the vector-valued Lagrange function defined for such a nonconvex optimization problem. Hence, Antczak proved the equivalence between an optimal solution and a saddle point of the scalar Lagrange function and the vector-valued Lagrange function in the considered semi-infinite minimax fractional programming problem under various (p, r)-invexity assumptions.

However, there are few papers on saddle point criteria for nonconvex minmax fractional programming problems in which functions involved are locally Lipschitz. In this paper, therefore, we consider a nondifferentiable minmax fractional programming problem with both inequality and equality constraints. For such nonsmooth optimization problem, we define seven Lagrange-type functions and, for each of them, we give a definition of a saddle point. Further, both parameter and parameter-free necessary optimality conditions are given for the considered minmax fractional programming problem. Then using the concepts of nondifferentiable (Ψ, Φ, ρ)-univexity and nondifferentiable (Ψ, Φ, ρ)-pseudounivexity introduced in the paper, we prove four sets of saddle point criteria for the nondifferentiable minmax fractional programming problem considered in the paper. Furthermore, we use these saddle point criteria to prove a variety of sufficient optimality conditions for the considered nondifferentiable minmax fractional programming problem. Thus, we prove the equivalence between a saddle point of each Lagrange function defined in the paper and an optimal solution in the considered nondifferentiable minmax fractional programming problem under the concepts of generalized convexity mentioned above.

2. Preliminaries

In this section, we provide some definitions and some results that we shall use in the sequel.

A function $f : X \to \mathbb{R}$ is locally Lipschitz near a point $\overline{x} \in X$ if there exist scalars $K > 0$ and $\varepsilon > 0$ such that, the following inequality $|f(y) - f(z)| \leqq K\|y - z\|$ holds for all y, $z \in \overline{x} + \varepsilon B$, where B signifies the open unit ball in X, so that $\overline{x} + \varepsilon B$ is the open ball of radius ε about $\overline{x}$.

Definition 1. *[15] The Clarke generalized directional derivative of a locally Lipschitz function $f : X \to R$ near $\overline{x} \in X$ in the direction $d \in X$, denoted $f^0(\overline{x}; d)$, is given by*

$$f^0(\overline{x}; d) = \limsup_{\substack{y \to \overline{x} \\ \theta \downarrow 0}} \frac{f(y + \theta d) - f(y)}{\theta}.$$

Let us denote by X^* the (topological) dual space of a Banach space X.

Definition 2. *[15] The Clarke generalized gradient of a locally Lipschitz function $f : X \to \mathbb{R}$ at $\overline{x} \in X$, denoted $\partial f(\overline{x})$, is defined as follows:*

$$\partial f(\overline{x}) = \left\{\xi \in X^* : f^0(\overline{x}; d) \geq \langle \xi, d \rangle \text{ for all } d \in X\right\}$$

We now introduce a concept of nondifferentiable (Ψ, Φ, ρ)-univexity as a generalization of the definition of a univex function given by Bector et al. [13] for differentiable optimization problems and the definition of a locally Lipschitz (Φ, ρ)-invex function introduced by Antczak and Stasiak [9].

Let $\overline{x} \in X$ be given and, moreover, assume that the function $f : X \to \mathbb{R}$ is Lipschitz near $\overline{x}$.

Definition 3. *The function f is said to be locally Lipschitz (strictly) (Ψ, Φ, ρ)-univex at $\overline{x}$ on X if there exist functions $\Psi : \mathbb{R} \to \mathbb{R}$, $\Phi : X \times X \times X^* \times \mathbb{R} \to \mathbb{R}$*

and $\rho \in \mathbb{R}$ *such that* $\Phi(x, \overline{x}; (\cdot, \cdot))$ *is convex for all* $x \in X$, $\Phi(x, \overline{x}; (0, a)) \geq 0$ *for all* $a \in \mathbb{R}_+$, *such that the inequality*

$$\Psi\left(f(x) - f(\overline{x})\right) \geq \Phi\left(x, \overline{x}; (\xi, \rho)\right) \;\; (>). \tag{1}$$

holds for any $\xi \in \partial f(\overline{x})$ *and all* $x \in X$, $(x \neq \overline{x})$*. If inequality (1) is satisfied at each* $\overline{x} \in X$*, then* f *is said to be locally Lipschitz (strictly)* (Ψ, Φ, ρ)*-univex on* X.

Further, we introduce the definition of nondifferentiable (Ψ, Φ, ρ)-pseudounivexity for a locally Lipschitz function.

Definition 4. *The function* f *is said to be locally Lipschitz (strictly)* (Ψ, Φ, ρ)*-pseudounivex at* $\overline{x}$ *on* X *if there exist functions* $\Psi : \mathbb{R} \to \mathbb{R}$, $\Phi : X \times X \times X^* \times \mathbb{R} \to \mathbb{R}$ *and* $\rho \in \mathbb{R}$ *such that* $\Phi(x, \overline{x}; (\cdot, \cdot))$ *is convex for all* $x \in X$, $\Phi(x, \overline{x}; (0, a)) \geq 0$ *for all* $a \in \mathbb{R}_+$ *such that the following relation*

$$\Phi\left(x, \overline{x}; (\xi, \rho)\right) \geq 0 \;\Rightarrow\; \Psi\left(f(x) - f(\overline{x})\right) \geq 0 \quad (>). \tag{2}$$

holds for any $\xi \in \partial f(\overline{x})$ *and all* $x \in X$, $(x \neq \overline{x})$*. The function* f *is said to be (strictly)* (Ψ, Φ, ρ)*-pseudounivex on* X *if it is (strictly)* (Ψ, Φ, ρ)*-pseudounivex at each* $\overline{x} \in X$, $(x \neq \overline{x})$.

In the proofs of the results established in the paper, we shall also make frequent use of the following auxiliary result.

Lemma 5. *[41] For each* $x \in X$,

$$\varphi(x) \equiv \max_{1 \leq i \leq p} \frac{f_i(x)}{g_i(x)} = \max_{\lambda \in \Lambda} \frac{\sum_{i=1}^{p} \lambda_i f_i(x)}{\sum_{i=1}^{p} \lambda_i g_i(x)},$$

where $\Lambda = \{\lambda = (\lambda_1, ..., \lambda_p) \in \mathbb{R}^p : \lambda_i \geq 0,\, i = 1, ..., p \,\wedge\, \sum_{i=1}^{p} \lambda_i = 1\}$.

In the remainder of this paper, we assume that the functions f_i, q_i, $i \in I$, and g_j, $j \in J$, h_k, $k \in K$, are locally Lipschitz on X.

We conclude this section by recalling both parametric and parameter-free necessary optimality conditions for the considered nonsmooth discrete minimax fractional programming problem (P). These results will be needed for proving various saddle point criteria. For the sake of brevity, we will henceforth refer to an optimal solution of problem (P) as normal if it satisfies an appropriate constraint qualification. First, we extend the so-called parametric necessary optimality conditions to the case of a nonsmooth discrete minimax fractional programming problem with both inequality and equality constraints (see, for example, [27], [31], [41]).

Theorem 6. *Let $\overline{x} \in D$ be a normal optimal solution in the considered nonsmooth discrete minimax fractional programming problem (P) and $\overline{v} = \varphi(\overline{x})$. Then there exist $\overline{\lambda} \in \Lambda$, $\overline{\mu} \in \mathbb{R}^m_+$ and $\overline{\vartheta} \in \mathbb{R}^r$ such that*

$$0 \in \sum_{i=1}^{p} \overline{\lambda}_i [\partial f_i(\overline{x}) - \overline{v} \partial g_i(\overline{x})] + \sum_{j=1}^{m} \overline{\mu}_j \partial g_j(\overline{x}) + \sum_{k=1}^{r} \overline{\vartheta}_k \partial h_k(\overline{x}), \tag{3}$$

$$\overline{\lambda}_i [f_i(\overline{x}) - \overline{v} q_i(\overline{x})] = 0, \quad i \in I, \tag{4}$$

$$\overline{\mu}_j g_j(\overline{x}) = 0, \quad j \in J. \tag{5}$$

It is easily seen that one obtains the following parameter-free version of Theorem 6 by eliminating the parameter $\overline{v}$ and redefining Lagrange multipliers associated with both inequality and equality constraints (see [41]).

Theorem 7. *Let $\overline{x}$ be a normal optimal solution in the considered nonsmooth discrete minimax fractional programming problem (P). Then there exist $\overline{\lambda} \in \Lambda$, $\overline{\mu} \in \mathbb{R}^m_+$ and $\overline{\vartheta} \in \mathbb{R}^r$ such that*

$$0 \in \sum_{i=1}^{p} \overline{\lambda}_i [M(\overline{x}, \overline{\lambda}) \partial f_i(\overline{x}) + N(\overline{x}, \overline{\lambda}) \partial(-q_i)(\overline{x})] + \sum_{j=1}^{m} \overline{\mu}_j \partial g_j(\overline{x}) + \sum_{k=1}^{r} \overline{\vartheta}_k \partial h_k(\overline{x}), \tag{6}$$

$$\overline{\lambda}_i [M(\overline{x}, \overline{\lambda}) f_i(\overline{x}) - N(\overline{x}, \overline{\lambda}) q_i(\overline{x})] = 0, \quad i \in I, \tag{7}$$

$$\max_{1 \leq i \leq p} \frac{f_i(\overline{x})}{q_i(\overline{x})} = \frac{N(\overline{x}, \overline{\lambda})}{M(\overline{x}, \overline{\lambda})} \tag{8}$$

$$\overline{\mu}_j g_j(\overline{x}) = 0, \quad j \in J, \tag{9}$$

where $N(\overline{x}, \overline{\lambda}) = \sum_{i=1}^{p} \overline{\lambda}_i f_i(\overline{x})$ and $M(\overline{x}, \overline{\lambda}) = \sum_{i=1}^{p} \overline{\lambda}_i q_i(\overline{x})$.

The theorem given above will be useful as a guide for formulating parameter-free saddle point criteria and sufficient optimality conditions and for the considered nonsmooth discrete minimax fractional programming problem (P).

3. Saddle-Point-Type Optimality Conditions

In this section, we present several sets of parametric and nonparametric saddle-point-type necessary and sufficient optimality conditions for the considered nonsmooth discrete minimax fractional programming problem (P). We state our first result in terms of the Lagrangian-type function $L_1 : X \times \mathbb{R}^m_+ \times \mathbb{R}^r \to \mathbb{R}$ defined, for fixed $v \in \mathbb{R}_+$, by

$$L_{1(v)}(x, \mu, \vartheta) = \max_{1 \leq i \leq p} [f_i(x) - vq_i(x)] + \sum_{j=1}^{m} \mu_j g_j(x) + \sum_{k=1}^{r} \vartheta_k h_k(x).$$

It is clear from Lemma 5 that this function can be expressed in the following equivalent form :

$$L_{1(\overline{\lambda},v)}(x, \mu, \vartheta) = \max_{\lambda \in \Lambda} \sum_{i=1}^{p} \lambda_i [f_i(x) - vq_i(x)] + \sum_{j=1}^{m} \mu_j g_j(x) + \sum_{k=1}^{r} \vartheta_k h_k(x),$$

where $\overline{\lambda} \in \Lambda$ is a vector such that $\sum_{i=1}^{p} \overline{\lambda}_i [f_i(x) - vq_i(x)] = \max_{\lambda \in \Lambda} \sum_{i=1}^{p} \lambda_i [f_i(x) - vq_i(x)]$ for a given $x \in X$.

Hence, one can use either one of these functions to state a set of optimality conditions for problem (P) by using of a saddle point criteria.

Theorem 8. *Let $\overline{x} \in D$ be a normal optimal solution of problem (P) and the necessary optimality conditions (3)-(5) be satisfied at $\overline{x}$ with Lagrange multipliers $\overline{\lambda} \in \Lambda$, $\overline{\mu} \in \mathbb{R}^m_+$ and $\overline{\vartheta} \in \mathbb{R}^r$. Further, assume that $\overline{v} = \varphi(\overline{x}) \geq 0$ and either one of the following two sets of hypotheses is satisfied:*

(a) (i) *for each $i \in I$, f_i is (Ψ, Φ, ρ_{f_i})-univex and $-q_i$ is (Ψ, Φ, ρ_{q_i})-univex at $\overline{x}$ on X;*

(ii) *for each $j \in J(\overline{x})$, g_j is (Ψ, Φ, ρ_{g_j})-univex at $\overline{x}$ on X;*

(iii) *for each $k \in K^+ = \{k \in K : \overline{\vartheta}_k > 0\}$, h_k is $(\Psi, \Phi, \rho^+_{h_k})$-univex at $\overline{x}$ on X;*

(iv) *for each $k \in K^- = \{k \in K : \overline{\vartheta}_k < 0\}$, $-h_k$ is $(\Psi, \Phi, \rho^-_{h_k})$-univex at $\overline{x}$ on X;*

(v) *the function Ψ is superlinear and $\Psi(a) \geq 0 \Rightarrow a \geq 0$;*

(vi) $\sum_{i=1}^{p} \overline{\lambda}_i [\rho_{f_i} + \overline{v} \rho_{q_i}] + \sum_{j=1}^{m} \overline{\mu}_j \rho_{h_j} + \sum_{k \in K^+} \overline{\vartheta}_k \rho^+_{h_k} - \sum_{k \in K^-} \overline{\vartheta}_k \rho^-_{h_k} \geq 0.$

(b) *(i) the function*

$$z \to \Gamma_{1(\overline{\lambda},\overline{v})}(z,\overline{\mu},\overline{\vartheta}) = \sum_{i=1}^{p} \overline{\lambda}_i[f_i(z) - \overline{v}q_i(z)] + \sum_{j=1}^{m} \overline{\mu}_j g_j(z) + \sum_{k=1}^{r} \overline{\vartheta}_k h_k(z)$$

is $(\Psi_{\Gamma_1}, \Phi_{\Gamma_1}, \rho_{\Gamma_1})$*-pseudounivex at* $\overline{x}$ *on* X*, where* Ψ_{Γ_1} *is an increasing function with* $\Psi_{\Gamma_1}(a) \geq 0 \Rightarrow a \geq 0$ *and* $\rho_{\Gamma_1} \geq 0$.
Then, the following inequalities

$$L_{1(\overline{\lambda},\overline{v})}(\overline{x},\mu,\overline{\vartheta}) \leq L_{1(\overline{\lambda},\overline{v})}(\overline{x},\overline{\mu},\overline{\vartheta}) \leq L_{1(\overline{\lambda},\overline{v})}(x,\overline{\mu},\overline{\vartheta}) \qquad (10)$$

hold for all $x \in X$ *and any* $\mu \in \mathbb{R}^m_+$.

Proof. By assumption, $\overline{x} \in D$ is a normal optimal solution in problem (P) and the parametric necessary optimality conditions (3)-(5) are fulfilled at this point with Lagrange multipliers $\overline{\lambda} \in \Lambda$, $\overline{\mu} \in \mathbb{R}^m_+$ and $\overline{\vartheta} \in \mathbb{R}^r$. Hence, by the parametric necessary optimality condition (3), there exist $\xi_i \in \partial f_i(\overline{x})$, $-\zeta_i \in \partial(-q_i)(\overline{x})$, $i \in I$, $\varsigma_j \in \partial g_j(\overline{x})$, $j \in J$, $\delta_k \in \partial h_k(\overline{x})$, $k \in K$, such that

$$\sum_{i=1}^{p} \overline{\lambda}_i(\xi_i - \overline{v}\zeta_i) + \sum_{j=1}^{m} \overline{\mu}_j\varsigma_j + \sum_{k=1}^{r} \overline{\vartheta}_k\delta_k = 0. \qquad (11)$$

Now, we prove the first inequality in (10), that is, we show that the inequality $L_{1(\overline{\lambda},\overline{v})}(\overline{x},\mu,\overline{\vartheta}) \leq L_{1(\overline{\lambda},\overline{v})}(\overline{x},\overline{\mu},\overline{\vartheta})$ is satisfied for any $\mu \in \mathbb{R}^m_+$. Using the parametric necessary optimality condition (5) together with $\mu \in \mathbb{R}^m_+$ and $\overline{x} \in D$, we get

$$\sum_{j=1}^{m} \mu_j g_j(\overline{x}) \leq \sum_{j=1}^{m} \overline{\mu}_j g_j(\overline{x}). \qquad (12)$$

Hence, (12) implies that, for any $\mu \in \mathbb{R}^m_+$,

$$\begin{array}{c} \max_{1\leq i\leq p} \sum_{i=1}^{p} \overline{\lambda}_i[f_i(\overline{x}) - \overline{v}q_i(\overline{x})] + \sum_{j=1}^{m} \mu_j g_j(\overline{x}) + \sum_{k=1}^{r} \overline{\vartheta}_k h_j(\overline{x}) \leq \\ \max_{1\leq i\leq p} \sum_{i=1}^{p} \overline{\lambda}_i[f_i(\overline{x}) - \overline{v}q_i(\overline{x})] + \sum_{j=1}^{m} \overline{\mu}_j g_j(\overline{x}) + \sum_{k=1}^{r} \overline{\vartheta}_k h_j(\overline{x}) \end{array}.$$

By the definition of the Lagrange function L_1, we get the first inequality in (10).

Now, we prove the second inequality in (10). Let x be an arbitrary point of X. Using hypotheses specified in (i)-(iv), by Definition 3, it follows that the following inequalities

$$\Psi(f_i(x) - f_i(\overline{x}))) \geq \Phi(x,\overline{x};(\xi_i,\rho_{f_i}(x,\overline{x})),\ i \in I, \qquad (13)$$

$$\Psi(-q_i(x) + q_i(\overline{x}))) \geq \Phi(x, \overline{x}; (-\zeta_i, \rho_{q_i}(x, \overline{x})), \ i \in I, \tag{14}$$

$$\Psi(g_j(x) - g_j(\overline{x}))) \geq \Phi(x, \overline{x}; (\varsigma_j, \rho_{g_j}(x, \overline{x})), \ j \in J. \tag{15}$$

$$\Psi(h_k(x) - h_k(\overline{x}))) \geq \Phi(x, \overline{x}; (\delta_k, \rho^+_{h_k}(x, \overline{x})), \ k \in K^+, \tag{16}$$

$$\Psi(-h_k(x) + h_k(\overline{x}))) \geq \Phi(x, \overline{x}; (-\delta_k, \rho^-_{h_k}(x, \overline{x})), \ k \in K^- \tag{17}$$

hold for all $x \in X$. Multiplying (13) by $\overline{\lambda}_i$ and summing over $i \in I$, (14) by $\overline{v}\overline{\lambda}_i$ and summing over $i \in I$, (15) by $\overline{\mu}_j$ and summing over $j \in J$, (16) by $\overline{\vartheta}_k$ and summing over $k \in K^+$, (17) by $-\overline{\vartheta}_k$ and summing over $k \in K^-$, and finally, adding the resulting inequalities, using the superlinearity of Ψ and (12), we get

$$\begin{gathered}\Psi\Big(\textstyle\sum_{i=1}^{p}\overline{\lambda}_i[f_i(x) - \overline{v}q_i(x)] + \sum_{j=1}^{m}\overline{\mu}_j g_j(x) + \sum_{k\in K^+}\overline{\vartheta}_k h_k(x) + \\ \textstyle\sum_{k\in K^-}\overline{\vartheta}_k h_k(x) - \sum_{i=1}^{p}\overline{\lambda}_i[f_i(\overline{x}) - \overline{v}q_i(\overline{x})] - \sum_{j=1}^{m}\overline{\mu}_j g_j(\overline{x}) - \\ \textstyle\sum_{k\in K^+}\overline{\vartheta}_k h_k(\overline{x}) - \sum_{k\in K^-}\overline{\vartheta}_k h_k(\overline{x})\Big) \geq \sum_{i=1}^{p}\overline{\lambda}_i\Phi(x, \overline{x}; (\xi_i, \rho_{f_i})) + \\ \textstyle\sum_{i=1}^{p}\overline{v}\overline{\lambda}_i\Phi(x, \overline{x}; (-\zeta_i, \rho_{q_i})) + \sum_{j=1}^{m}\overline{\mu}_j\Phi(x, \overline{x}; (\varsigma_j, \rho_{g_j}) + \\ \textstyle\sum_{k\in K^+}\overline{\vartheta}_k\Phi(x, \overline{x}; (\delta_k, \rho^+_{h_k}) - \sum_{k\in K^-}\overline{\vartheta}_k\Phi(x, \overline{x}; (-\delta_k, \rho^-_{h_k}).\end{gathered} \tag{18}$$

Letting

$$w = \sum_{i=1}^{p}\overline{\lambda}_i(1 + \overline{v}) + \sum_{j=1}^{m}\overline{\mu}_j + \sum_{k\in K^+}\overline{\vartheta}_k - \sum_{k\in K^-}\overline{\vartheta}_k, \tag{19}$$

and defining

$$\overline{\alpha}_i = \frac{\overline{\lambda}_i}{w}, \ i \in I, \ \overline{\beta}_j = \frac{\overline{\mu}_j}{w}, \ j \in J, \overline{\gamma}_k = \frac{\overline{\vartheta}_k}{w}, k \in K^+, \overline{\gamma}_k = \frac{-\overline{\vartheta}_k}{w}, k \in K^-, \tag{20}$$

we see that

$$w > 0, 0 \leq \overline{\alpha}_i \leq 1, i \in I, 0 \leq \overline{\beta}_j \leq 1, j \in J, 0 \leq \overline{\gamma}_k \leq 1, k \in K^+ \cup K^- \tag{21}$$

and, moreover,

$$\sum_{i=1}^{p}\overline{\alpha}_i(1 + \overline{v}) + \sum_{j=1}^{m}\overline{\beta}_j + \sum_{k\in K^+}\overline{\gamma}_k + \sum_{k\in K^-}\overline{\gamma}_k = 1. \tag{22}$$

Now, dividing (18) by $w > 0$, using (19)-(20) together with the superlinearity of Ψ, we get

$$\Psi\Big(\sum_{i=1}^{p}\overline{\alpha}_i[f_i(x)-\overline{v}q_i(x)]+\sum_{j=1}^{m}\overline{\beta}_j g_j(x)+\sum_{k\in K^+}\overline{\gamma}_k h_k(x)-$$
$$\sum_{k\in K^-}\overline{\gamma}_k h_k(x)-\sum_{i=1}^{p}\overline{\alpha}_i[f_i(\overline{x})-\overline{v}q_i(\overline{x})]-\sum_{j=1}^{m}\overline{\beta}_j g_j(\overline{x})-$$
$$\sum_{k\in K^+}\overline{\gamma}_k h_k(\overline{x})+\sum_{k\in K^-}\overline{\gamma}_k h_k(\overline{x})\Big)\geq\sum_{i=1}^{p}\overline{\alpha}_i\Phi(x,\overline{x};(\xi_i,\rho_{f_i}))+$$
$$\sum_{i=1}^{p}\overline{v}\overline{\alpha}_i\Phi(x,\overline{x};(-\zeta_i,\rho_{q_i}))+\sum_{j=1}^{m}\overline{\beta}_j\Phi(x,\overline{x};(\varsigma_j,\rho_{g_j}))+$$
$$\sum_{k\in K^+}\overline{\gamma}_k\Phi(x,\overline{x};(\delta_k,\rho_{h_k}^{+})+\sum_{k\in K^-}\overline{\gamma}_k\Phi(x,\overline{x};(-\delta_k,\rho_{h_k}^{-}).$$

Using convexity of the function $\Phi(x,\overline{x};(\cdot,\cdot))$ together with (19)-(20), the above inequality yields

$$\Psi\Big(\sum_{i=1}^{p}\overline{\alpha}_i[f_i(x)-\overline{v}q_i(x)]+\sum_{j=1}^{m}\overline{\beta}_j g_j(x)+\sum_{k\in K^+}\overline{\gamma}_k h_k(x)-$$
$$\sum_{k\in K^-}\overline{\gamma}_k h_k(x)-\sum_{i=1}^{p}\overline{\alpha}_i[f_i(\overline{x})-\overline{v}q_i(\overline{x})]-\sum_{j=1}^{m}\overline{\beta}_j g_j(\overline{x})-$$
$$\sum_{k\in K^+}\overline{\gamma}_k h_k(\overline{x})+\sum_{k\in K^-}\overline{\gamma}_k h_k(\overline{x})\Big)\geq$$
$$\Phi\Big(x,\overline{x};\Big(\sum_{i=1}^{p}\overline{\alpha}_i(\xi_i-\overline{v}\zeta_i)+\sum_{j=1}^{m}\overline{\beta}_j\varsigma_j+$$
$$\sum_{k\in K^+}\overline{\gamma}_k\delta_k-\sum_{k\in K^-}\overline{\gamma}_k\delta_k,\sum_{i=1}^{p}\overline{\alpha}_i\left(\rho_{f_i}+\overline{v}\rho_{q_i}\right)+\sum_{j=1}^{m}\overline{\beta}_j\rho_{g_j}+$$
$$\sum_{k\in K^+}\overline{\gamma}_k\rho_{h_k}^{+}+\sum_{k\in K^-}\overline{\gamma}_k\rho_{h_k}^{-}\Big)\Big).$$

By (19)-(20), the above inequality gives

$$\Psi\Big(\frac{1}{w}\Big(\sum_{i=1}^{p}\overline{\lambda}_i[f_i(x)-\overline{v}q_i(x)]+\sum_{j=1}^{m}\overline{\mu}_j g_j(x)+\sum_{k\in K^+}\overline{\vartheta}_k h_k(x)+$$
$$\sum_{k\in K^-}\overline{\vartheta}_k h_k(x)-\sum_{i=1}^{p}\overline{\lambda}_i[f_i(\overline{x})-\overline{v}q_i(\overline{x})]-\sum_{j=1}^{m}\overline{\mu}_j g_j(\overline{x})-$$
$$\sum_{k\in K^+}\overline{\vartheta}_k h_k(\overline{x})-\sum_{k\in K^-}\overline{\vartheta}_k h_k(\overline{x})\Big)\geq$$
$$\Phi\Big(x,\overline{x};\frac{1}{w}\Big(\sum_{i=1}^{p}\overline{\lambda}_i(\xi_i-\overline{v}\zeta_i)+\sum_{j=1}^{m}\overline{\mu}_j\varsigma_j+\sum_{k=1}^{r}\overline{\vartheta}_k\delta_k\,,$$
$$\sum_{i=1}^{p}\overline{\lambda}_i\left(\rho_{f_i}+\overline{v}\rho_{q_i}\right)+\sum_{j=1}^{m}\overline{\mu}_j\rho_{g_j}+\sum_{k\in K^+}\overline{\vartheta}_k\rho_{h_k}^{+}-\sum_{k\in K^-}\overline{\vartheta}_k\rho_{h_k}^{-}\Big)\Big).$$

By (11) and since $\Phi\left(x,\overline{x};(0,a)\right)\geq 0$ for all $a\in\mathbb{R}_+$, by hypothesis (vi), it follows that

$$\Psi\Big(\tfrac{1}{w}\Big(\sum_{i=1}^{p}\overline{\lambda}_i[f_i(x)-\overline{v}q_i(x)]+\sum_{j=1}^{m}\overline{\mu}_j g_j(x)+$$
$$\sum_{k\in K^+}\overline{\vartheta}_k h_k(x)+\sum_{k\in K^-}\overline{\vartheta}_k h_k(x)-\sum_{i=1}^{p}\overline{\lambda}_i[f_i(\overline{x})-\overline{v}q_i(\overline{x})]-$$
$$\sum_{j=1}^{m}\overline{\mu}_j g_j(\overline{x})-\sum_{k\in K^+}\overline{\vartheta}_k h_k(\overline{x})-\sum_{k\in K^-}\overline{\vartheta}_k h_k(\overline{x})\Big)\geq 0.$$

By $\Psi(a)\geq 0\Rightarrow a\geq 0$, the above inequality implies

$$\sum_{i=1}^{p}\overline{\lambda}_i[f_i(x)-\overline{v}q_i(x)]+\sum_{j=1}^{m}\overline{\mu}_j g_j(x)+\sum_{k\in K^+}\overline{\vartheta}_k h_k(x)+$$
$$\sum_{k\in K^-}\overline{\vartheta}_k h_k(x)-\sum_{i=1}^{p}\overline{\lambda}_i[f_i(\overline{x})-\overline{v}q_i(\overline{x})]- \tag{23}$$
$$\sum_{j=1}^{m}\overline{\mu}_j g_j(\overline{x})-\sum_{k\in K^+}\overline{\vartheta}_k h_k(\overline{x})-\sum_{k\in K^-}\overline{\vartheta}_k h_k(\overline{x})\geq 0.$$

Hence, by Lemma 5, it follows that

$$\max_{1\leq i\leq p}[f_i(x)-\overline{v}q_i(x)]=\max_{\lambda\in\Lambda}\sum_{i=1}^{p}\lambda_i[f_i(x)-\overline{v}q_i(x)]\geq\sum_{i=1}^{p}\overline{\lambda}_i[f_i(x)-\overline{v}q_i(x)]. \tag{24}$$

By the necessary optimality condition (4), (23) and (24) yield

$$\max_{1\leq i\leq p}\sum_{i=1}^{p}\overline{\lambda}_i[f_i(x)-\overline{v}q_i(x)]+\sum_{j=1}^{m}\overline{\mu}_j g_j(x)+\sum_{k=1}^{r}\overline{\vartheta}_k h_k(x)\geq$$
$$\max_{1\leq i\leq p}\sum_{i=1}^{p}\overline{\lambda}_i[f_i(\overline{x})-\overline{v}q_i(\overline{x})]+\sum_{j=1}^{m}\overline{\mu}_j g_j(\overline{x})+\sum_{k=1}^{r}\overline{\vartheta}_k h_k(\overline{x}).$$

Hence, by the definition of the Lagrange function $L_{1(\overline{\lambda},\overline{v})}$, we have that, for all $x\in X$,

$$L_{1(\overline{\lambda},\overline{v})}(x,\overline{\mu})\geq L_{1(\overline{\lambda},\overline{v})}(\overline{x},\overline{\mu}).$$

Inasmuch as the first inequality of (10) has already been established, the proof of the theorem under hypothesis (a) is complete.

(b) By assumption, the function $z\to\Gamma_{1(\overline{\lambda},\overline{v})}(z,\overline{\mu},\overline{\vartheta})=\sum_{i=1}^{p}\overline{\lambda}_i[f_i(z)-\overline{v}q_i(z)]+\sum_{j=1}^{m}\overline{\mu}_j g_j(z)+\sum_{k=1}^{r}\overline{\vartheta}_k h_k(z)$ is $(\Psi_{\Gamma_1},\Phi_{\Gamma_1},\rho_{\Gamma_1})$-pseudounivex at $\overline{x}$ on X with $\rho_{\Gamma_1}\geq 0$. By assumption, $\overline{x}\in D$ is a normal optimal solution in problem (P). Thus, using (11) together with $\Phi_{\Gamma_1}(x,\overline{x};(0,a))\geq 0$ for all $a\in\mathbb{R}_+$, by Definition 4, we get that the following inequality

$$\Phi_{\Gamma_1}\big(x,\overline{x};\big(\sum_{i=1}^{p}\overline{\lambda}_i(\xi_i-\overline{v}\zeta_i)+\sum_{j=1}^{m}\overline{\mu}_j\varsigma_j+\sum_{k=1}^{r}\overline{\vartheta}_k\delta_k,\rho_{\Gamma_1}\big)\big)\geq 0$$

holds for all $x \in X$. Hence, by $(\Psi_{\Gamma_1}, \Phi_{\Gamma_1}, \rho_{\Gamma_1})$-pseudounivexity hypothesis (see Definition 4), the above inequality implies that the following inequality

$$\Psi_{\Gamma_1}\left(\Gamma_{1(\overline{\lambda},\overline{v})}(x, \overline{\mu}, \overline{\vartheta}) - \Gamma_{1(\overline{\lambda},\overline{v})}(\overline{x}, \overline{\mu}, \overline{\vartheta})\right) \geq 0$$

holds for all $x \in X$. Since $\Psi(a) \geq 0 \Rightarrow a \geq 0$, therefore, the inequality above gives

$$\Gamma_{1(\overline{\lambda},\overline{v})}(x, \overline{\mu}, \overline{\vartheta}) \geq \Gamma_{1(\overline{\lambda},\overline{v})}(\overline{x}, \overline{\mu}, \overline{\vartheta}).$$

Taking into account the fact that the right-hand side of this inequality is equal to zero and using (4) and (5), the above inequality reduces to (23). Hence, the rest of the proof is, therefore, identical to that of part (a). □

We next show that the inequalities in (10) constitute sufficient conditions for optimality of $\overline{x}$ without any feasibility assumption, constraint qualification, generalized convexity hypotheses, or complementary slackness condition.

Theorem 9. *Let $\overline{x} \in D$ and $\overline{v} = \varphi(\overline{x})$. Further, assume that there exist $\overline{\mu} \in \mathbb{R}^m_+$ and $\overline{\vartheta} \in \mathbb{R}^r$ such that the inequalities (10) hold for all $x \in X$ and $\mu \in \mathbb{R}^m_+$. Then $\sum_{j=1}^m \overline{\mu}_j g_j(\overline{x}) = 0$ and $\overline{x}$ is an optimal solution in the considered nonsmooth discrete minimax fractional programming problem (P).*

Proof. By $\overline{x} \in D$ and $\overline{\mu} \in \mathbb{R}^m_+$, it follows that

$$\sum_{j=1}^{m} \overline{\mu}_j g_j(\overline{x}) \leq 0. \tag{25}$$

From the first inequality of (10), we have that the following inequality $\sum_{j=1}^m (\mu_j - \overline{\mu}_j) g_j(\overline{x}) \leq 0$ holds for any $\mu \in \mathbb{R}^m_+$. Therefore, if we set $\mu_j = 0$ for $j \in J$, then we conclude that

$$\sum_{j=1}^{m} \overline{\mu}_j g_j(\overline{x}) \geq 0. \tag{26}$$

Combining (25) and (26), we get that

$$\sum_{j=1}^{m} \overline{\mu}_j g_j(\overline{x}) = 0. \tag{27}$$

In order to show that $\overline{x}$ is optimal in problem (P), let x be an arbitrary feasible solution of problem (P). Then, by $x \in D$, $\overline{\mu} \in \mathbb{R}^m_+$ and (27), the second inequality of (10) reduces to

$$\max_{1 \leq i \leq p} [f_i(\overline{x}) - \overline{v} q_i(\overline{x})] \leq \max_{1 \leq i \leq p} [f_i(x) - \overline{v} q_i(x)].$$

By the above inequality, it follows that $\overline{x}$ is an optimal solution in problem $(P\overline{v})$, that is, in the auxiliary parametric optimization problem defined in the Dinkelbach-type indirect approach to optimality in problem (P) (see [16] and also [12]). Since

$$\max_{1 \leq i \leq p} [f_i(\overline{x}) - \overline{v} q_i(\overline{x})] = 0 \ \Leftrightarrow \ \overline{v} = \varphi(\overline{x}),$$

by Lemma 3.1 [12], it follows that $\varphi(\overline{x}) = 0$, and hence, $\overline{x}$ is an optimal solution in the considered nonsmooth discrete minimax fractional programming problem (P). □

Next, we use a slightly different Lagrangian-type function $L_{2(\bar{x},\overline{\lambda})} : X \times \mathbb{R}^m_+ \times \mathbb{R}^r \to \mathbb{R}$ defined, for fixed $\bar{x} \in X$ and $\overline{\lambda} \in \Lambda$, by

$$L_{2(\bar{x},\overline{\lambda})}(x, \mu, \vartheta) = \max\nolimits_{1 \leq i \leq p} [M\left(\bar{x}, \overline{\lambda}\right) f_i(x) - N\left(\bar{x}, \overline{\lambda}\right) q_i(x)] + \sum\nolimits_{j=1}^m \mu_j g_j(x) + \sum\nolimits_{k=1}^r \vartheta_k h_k(x),$$

where M and N are as defined in Theorem 7. It is easily seen from Lemma 5 that

$$L_{2(\bar{x},\overline{\lambda})}(x, \mu, \vartheta) = \max\nolimits_{\lambda \in \Lambda} \sum\nolimits_{i=1}^p \lambda_i [M\left(\bar{x}, \overline{\lambda}\right) f_i(x) - N\left(\bar{x}, \overline{\lambda}\right) q_i(x)] + \sum\nolimits_{j=1}^m \mu_j g_j(x) + \sum\nolimits_{k=1}^r \vartheta_k h_k(x).$$

Theorem 10. *Let $\overline{x} \in D$ be a normal optimal solution of the considered nonsmooth discrete minimax fractional programming problem (P) with the associated vectors $\overline{\lambda}$, $\overline{\mu}$ and $\overline{\vartheta}$ specified in Theorem 7 and let $\overline{v} = \varphi(\overline{x})$. Further, assume that either one of the following two sets of hypotheses is satisfied:*

(a) *(i) for each $i \in I$, f_i is (Ψ, Φ, ρ_{f_i})-univex and $-q_i$ is (Ψ, Φ, ρ_{q_i})-univex at $\overline{x}$ on X;*

(ii) for each $j \in J$, g_j is (Ψ, Φ, ρ_{g_j})-univex at $\overline{x}$ on X;

(iii) for each $k \in K^+ = \left\{k \in K : \overline{\vartheta}_k > 0\right\}$, h_k *is* $(\Psi, \Phi, \rho^+_{h_k})$*-univex at* $\overline{x}$ *on* X*;*

(iv) for each $k \in K^- = \left\{k \in K : \overline{\vartheta}_k < 0\right\}$, $-h_k$ *is* $(\Psi, \Phi, \rho^-_{h_k})$*-univex at* $\overline{x}$ *on* X*;*

(v) the function Ψ *is superlinear and* $\Psi(a) \geq 0 \Rightarrow a \geq 0$*;*

(vi) $\sum_{i=1}^{p} \overline{\lambda}_i[M(\overline{x}, \overline{\lambda})\rho_{f_i} + N(\overline{x}, \overline{\lambda})\rho_{q_i}] + \sum_{j=1}^{m} \overline{\mu}_j \rho_{g_j} + \sum_{k\in K^+} \overline{\vartheta}_k \rho^+_{h_k} - \sum_{k\in K^-} \overline{\vartheta}_k \rho^-_{h_k} \geq 0.$

(b) (i) the function

$$z \to \Gamma_{2(\overline{x},\overline{\lambda})}(z, \overline{\mu}, \overline{\vartheta}) = \sum_{i=1}^{p} \overline{\lambda}_i[M(\overline{x}, \overline{\lambda}) f_i(z) - N(\overline{x}, \overline{\lambda}) q_i(z)] + \sum_{j=1}^{m} \overline{\mu}_j g_j(z) + \sum_{k=1}^{r} \overline{\vartheta}_k h_k(z)$$

$(\Psi_{\Gamma_2}, \Phi_{\Gamma_2}, \rho_{\Gamma_2})$*-pseudounivex at* $\overline{x}$ *on* X*, where* Ψ_{Γ_2} *is an increasing function with* $\Psi_{\Gamma_2}(a) \geq 0 \Rightarrow a \geq 0$ *and* $\rho_{\Gamma_2} \geq 0$.

Then

$$L_{2(\bar{x},\overline{\lambda})}(\overline{x}, \mu, \overline{\vartheta}) \leq L_{2(\bar{x},\overline{\lambda})}(\overline{x}, \overline{\mu}, \overline{\vartheta}) \leq L_{2(\bar{x},\overline{\lambda})}(x, \overline{\mu}, \overline{\vartheta}) \text{ for all } x \in X,\ \mu \in \mathbb{R}^m_+. \tag{28}$$

Proof. Since $\overline{x}$ is a normal optimal solution in problem (P), conditions (3)-(5) are fulfilled. By (3), there exist $\xi_i \in \partial f_i(\overline{x})$, $\zeta_i \in \partial(-q_i)(\overline{x})$, $i \in I$, and $\varsigma_j \in \partial g_j(\overline{x})$, $j \in J$, $\delta_k \in \partial h_k(\overline{x})$, $k \in K$, such that

$$\sum_{i=1}^{p} \overline{\lambda}_i(\xi_i - \overline{v}\zeta_i) + \sum_{j=1}^{m} \overline{\mu}_j \varsigma_j + \sum_{k=1}^{r} \overline{\vartheta}_k \delta_k = 0. \tag{29}$$

Let x be an arbitrary point of X. It follows from the assumptions specified in (i)-(iv) that, by Definition 3, inequalities (13)-(17) are satisfied. Since $\overline{\lambda} \geq 0$, $\overline{\mu}_j \geq 0$, $N(\overline{x}, \overline{\lambda}) \geq 0$, $M\left(\overline{x}, \overline{\lambda}\right) > 0$ and Ψ is superlinear, we deduce from these inequalities that

$$\begin{aligned}\Psi\Big(&\sum_{i=1}^{p} \overline{\lambda}_i M(\overline{x}, \overline{\lambda})[f_i(x) - f_i(\overline{x})] + \sum_{i=1}^{p} \overline{\lambda}_i N(\overline{x}, \overline{\lambda})[-q_i(x) + q_i(\overline{x})] \\ &+ \sum_{j=1}^{m} \overline{\mu}_j[g_j(x) - g_j(\overline{x})] + \sum_{k=1}^{r} \overline{\vartheta}_k\left[h_k(x) - h_k(\overline{x})\right]\Big) \geq \\ &\sum_{i=1}^{p} \overline{\lambda}_i M(\overline{x}, \overline{\lambda})\Phi(x, \overline{x}; (\xi_i, \rho_{f_i})) + N(\overline{x}, \overline{\lambda}) \sum_{i=1}^{p} \overline{v}\overline{\lambda}_i \Phi(x, \overline{x}; (-\zeta_i, \rho_{q_i})) \\ &+ \sum_{j=1}^{m} \overline{\mu}_j \Phi(x, \overline{x}; (\varsigma_j, \rho_{g_j})) + \sum_{k\in K^+} \overline{\vartheta}_k \Phi(x, \overline{x}; (\delta_k, \rho^+_{h_k}) - \\ &\sum_{k\in K^-} \overline{\vartheta}_k \Phi(x, \overline{x}; (-\delta_k, \rho^-_{h_k}).\end{aligned} \tag{30}$$

Let us denote

$$t=\sum_{i=1}^{p}\overline{\lambda}_i[M(\overline{x},\overline{\lambda})+N(\overline{x},\overline{\lambda})]+\sum_{j=1}^{m}\overline{\mu}_j+\sum_{k\in K^+}\overline{\vartheta}_k-\sum_{k\in K^-}\overline{\vartheta}_k. \tag{31}$$

Then, let us define

$$\begin{gathered}\overline{\alpha}_i^M=\tfrac{M(\overline{x},\overline{\lambda})\overline{\lambda}_i}{t},\ \overline{\alpha}_i^N=\tfrac{N(\overline{x},\overline{\lambda})\overline{\lambda}_i}{t}, i\in I, \overline{\beta}_j=\tfrac{\overline{\mu}_j}{t},\ j\in J,\\ \overline{\gamma}_k=\tfrac{\overline{\vartheta}_k}{t}, k\in K^+, \overline{\gamma}_k=\tfrac{-\overline{\vartheta}_k}{t}, k\in K^-.\end{gathered} \tag{32}$$

Hence, note that

$$t>0, 0\le\overline{\alpha}_i^N\le 1, 0\le\overline{\alpha}_i^M\le 1, i\in I, \tag{33}$$

$$0\le\overline{\beta}_j\le 1, j\in J, 0\le\overline{\gamma}_k\le 1, k\in K^+\cup K^-$$

and, moreover,

$$\sum_{i=1}^{p}(\overline{\alpha}_i^M+\overline{\alpha}_i^N)+\sum_{j=1}^{m}\overline{\beta}_j+\sum_{k\in K^+}\overline{\gamma}_k+\sum_{k\in K^-}\overline{\gamma}_k=1. \tag{34}$$

Now, dividing (30) by t and then using the superlinearity of Ψ and (32), we get

$$\begin{gathered}\textstyle\sum_{i=1}^{p}\overline{\alpha}_i^M\Phi(x,\overline{x};(\xi_i,\rho_{f_i}))+\sum_{i=1}^{p}\overline{\alpha}_i^N\Phi(x,\overline{x};(-\zeta_i,\rho_{q_i}))+\\ \textstyle\sum_{j=1}^{m}\overline{\beta}_j\Phi(x,\overline{x};(\varsigma_j,\rho_{g_j}))+\sum_{k\in K^+}\overline{\gamma}_k\Phi(x,\overline{x};(\delta_k,\rho_{h_k}^+))+\\ \textstyle\sum_{k\in K^-}\overline{\gamma}_k\Phi(x,\overline{x};(-\delta_k,\rho_{h_k}^-))\le\Psi\left(\sum_{i=1}^{p}[\overline{\alpha}_i^M f_i(x)-\overline{\alpha}_i^N g_i(x)]+\right.\\ \left.\textstyle\sum_{j=1}^{m}\overline{\beta}_j g_j(x)+\sum_{k\in K^+}\overline{\gamma}_k h_k(x)-\sum_{k\in K^-}\overline{\gamma}_k h_k(x)\right).\end{gathered}$$

Since (33) and (34) are satisfied, using convexity of the function $\Phi(x,\overline{x};(\cdot,\cdot))$, the above inequality yields

$$\begin{gathered}\textstyle\Phi\Big(x,\overline{x};\Big(\sum_{i=1}^{p}\left(\overline{\alpha}_i^M\xi_i-\overline{\alpha}_i^N\zeta_i\right)+\sum_{j=1}^{m}\overline{\beta}_j\varsigma_j+\\ \textstyle\sum_{k\in K^+}\overline{\gamma}_k\delta_k-\sum_{k\in K^-}\overline{\gamma}_k\delta_k\ ,\ \sum_{i=1}^{p}[\overline{\alpha}_i^M\rho_{f_i}+\overline{\alpha}_i^N\rho_{q_i}]+\\ \textstyle\sum_{j=1}^{m}\overline{\beta}_j\rho_{g_j}+\sum_{k\in K^+}\overline{\gamma}_k\rho_{h_k}^+ +\sum_{k\in K^-}\overline{\gamma}_k\rho_{h_k}^-\Big)\Big)\le\\ \textstyle\Psi\left(\sum_{i=1}^{p}[\overline{\alpha}_i^M f_i(x)-\overline{\alpha}_i^N g_i(x)]+\sum_{j=1}^{m}\overline{\beta}_j g_j(x)+\right.\\ \left.\textstyle\sum_{k\in K^+}\overline{\gamma}_k h_k(x)+\sum_{k\in K^-}\overline{\gamma}_k h_k(x)\right).\end{gathered}$$

Hence, by (32), we have

$$\Phi\Big(x, \overline{x}; \tfrac{1}{t}\Big(\sum_{i=1}^{p} \overline{\lambda}_i \left(M(\overline{x}, \overline{\lambda})\xi_i - N(\overline{x}, \overline{\lambda})\zeta_i\right) + \sum_{j=1}^{m} \overline{\mu}_j \varsigma_j +$$
$$\sum_{k\in K^+} \overline{\vartheta}_k \delta_k + \sum_{k\in K^-} \overline{\vartheta}_k \delta_k, \sum_{i=1}^{p} \overline{\lambda}_i [M(\overline{x}, \overline{\lambda})\rho_{f_i} + N(\overline{x}, \overline{\lambda})\rho_{q_i}] +$$
$$\sum_{j=1}^{m} \overline{\mu}_j \rho_{g_j} + \sum_{k\in K^+} \overline{\vartheta}_k \rho_{h_k}^{+} - \sum_{k\in K^-} \overline{\vartheta}_k \rho_{h_k}^{-} \Big)\Big) \leq$$
$$\Psi\Big(\tfrac{1}{t}\Big(\sum_{i=1}^{p} \overline{\lambda}_i [M(\overline{x}, \overline{\lambda}) f_i(x) - N(\overline{x}, \overline{\lambda}) q_i(x)] +$$
$$\sum_{j=1}^{m} \overline{\mu}_j g_j(x) + \sum_{k=1}^{r} \overline{\vartheta}_k h_k(x) \Big)\Big).$$

By Definition 3, it follows that $\Phi\left(x, \overline{x}; (0, a)\right) \geq 0$ for all $a \in \mathbb{R}_+$. Thus, using the necessary optimality condition (6) and hypothesis (vi), we get

$$\Psi\Big(\tfrac{1}{t}\Big(\sum_{i=1}^{p} \overline{\lambda}_i [M(\overline{x}, \overline{\lambda}) f_i(x) - N(\overline{x}, \overline{\lambda}) q_i(x)] +$$
$$\sum_{j=1}^{m} \overline{\mu}_j g_j(x) + \sum_{k=1}^{r} \overline{\vartheta}_k h_k(x) \Big)\Big) \geq 0.$$

Since $\Psi(a) \geq 0 \Rightarrow a \geq 0$, therefore, the above inequality gives

$$\sum_{i=1}^{p} \overline{\lambda}_i [M(\overline{x}, \overline{\lambda}) f_i(x) - N(\overline{x}, \overline{\lambda}) q_i(x)] + \sum_{j=1}^{m} \overline{\mu}_j g_j(x) + \sum_{k=1}^{r} \overline{\vartheta}_k h_k(x) \geq 0. \quad (35)$$

Now, using Lemma 5, we have

$$\max_{1\leq i\leq p} [M(\overline{x}, \overline{\lambda}) f_i(x) - N(\overline{x}, \overline{\lambda}) q_i(x)] =$$
$$\max_{\lambda\in\Lambda} \sum_{i=1}^{p} \lambda_i [M(\overline{x}, \overline{\lambda}) f_i(x) - N(\overline{x}, \overline{\lambda}) q_i(x)] \geq \quad (36)$$
$$\sum_{i=1}^{p} \overline{\lambda}_i [M(\overline{x}, \overline{\lambda}) f_i(x) - N(\overline{x}, \overline{\lambda}) q_i(x)].$$

Combining (35) and (36), we get

$$\max_{1\leq i\leq p} [M(\overline{x}, \overline{\lambda}) f_i(x) - N(\overline{x}, \overline{\lambda}) q_i(x)] + \sum_{j=1}^{m} \overline{\mu}_j g_j(x) + \sum_{k=1}^{r} \overline{\vartheta}_k h_k(x) \geq 0.$$

By the necessary optimality conditions (7), (9) and $\overline{x} \in D$, we conclude that

$$\max_{1\leq i\leq p} [M(\overline{x}, \overline{\lambda}) f_i(x) - N(\overline{x}, \overline{\lambda}) q_i(x)] + \sum_{j=1}^{m} \overline{\mu}_j g_j(x) + \sum_{k=1}^{r} \overline{\vartheta}_k h_k(x) \geq$$
$$\max_{1\leq i\leq p} [M(\overline{x}, \overline{\lambda}) f_i(\overline{x}) - N(\overline{x}, \overline{\lambda}) q_i(\overline{x})] + \sum_{j=1}^{m} \overline{\mu}_j g_j(\overline{x}) + \sum_{k=1}^{r} \overline{\vartheta}_k h_k(\overline{x}).$$

By the definition of the Lagrange function $L_{2(\overline{x},\overline{\lambda})}$, it follows that the second inequality of (28) is fulfilled for all $x \in X$.

Inasmuch as the first inequality of (28) holds trivially, the proof of the theorem under hypothesis (a) is completed.

b) By assumption, the function $z \rightarrow \Gamma_{2(\overline{x},\overline{\lambda})}(z,\overline{\mu},\overline{\vartheta}) = \sum_{i=1}^{p} \overline{\lambda}_i[M(\overline{x},\overline{\lambda})f_i(z) - N(\overline{x},\overline{\lambda})q_i(z)] + \sum_{j=1}^{m} \overline{\mu}_j g_j(z) + \sum_{k=1}^{r} \vartheta_k h_k(z)$ is $(\Psi_{\Gamma_2}, \Phi_{\Gamma_2}, \rho_{\Gamma_2})$-pseudounivex at $\overline{x}$ on X with $\rho \geq 0$. By assumption, $\overline{x} \in D$ is a normal optimal solution of problem (P). Hence, using the necessary optimality condition (6) together with $\rho_{\Gamma_2} \geq 0$ and $\Phi\left(x, \overline{x}; (0, a)\right) \geq 0$ for all $a \in \mathbb{R}_+$, we obtain that the following inequality

$$\Phi\big(x,\overline{x};\big(\sum_{i=1}^{p}\overline{\lambda}_i(M(\overline{x},\overline{\lambda})\xi_i - N(\overline{x},\overline{\lambda})\zeta_i) + \sum_{j=1}^{m}\overline{\mu}_j\varsigma_j, \rho_{\Gamma_2}\big)\big) \geq 0 \qquad (37)$$

holds for all $x \in X$. Hence, by $(\Psi_{\Gamma_2}, \Phi_{\Gamma_2}, \rho_{\Gamma_2})$-pseudounivexity hypothesis (see Definition 4), (37) implies that the following inequality

$$\Psi\big(\Gamma_{2(\overline{x},\overline{\lambda})}(x,\overline{\mu},\overline{\vartheta}) - \Gamma_{2(\overline{x},\overline{\lambda})}(\overline{x},\overline{\mu},\overline{\vartheta})\big) \geq 0$$

holds for all $x \in X$. Taking into account the assumption $\Psi(a) \geq 0 \Rightarrow a \geq 0$, we get that

$$\Gamma_{2(\overline{x},\overline{\lambda})}(x,\overline{\mu},\overline{\vartheta}) \geq \Gamma_{2(\overline{x},\overline{\lambda})}(\overline{x},\overline{\mu},\overline{\vartheta}).$$

Since the right-hand side of this inequality is equal to zero because of (4) and (5), it reduces to (23), and, therefore, the rest of the proof is identical to that of part (a). □

Theorem 11. *Let $\overline{x} \in D$ and $\varphi(\overline{x}) = N(\overline{x},\overline{\lambda})/M(\overline{x},\overline{\lambda})$ for some $\overline{\lambda} \in \Lambda$. Further, assume that there exist $\overline{\mu} \in \mathbb{R}_+^m$ and $\overline{\vartheta} \in \mathbb{R}^r$ such that inequalities (28) hold for all $x \in X$ and $\mu \in \mathbb{R}_+^m$. Then $\mu_j g_j(\overline{x}) = 0,\ j \in J$, and $\overline{x}$ is an optimal solution of the considered nonsmooth discrete minimax fractional programming problem (P).*

Proof. Proof is similar to that of Theorem 9. □

As a direct consequence of Theorems 9 and 11, it is possible to obtain another set of saddle-point-type optimality criteria for problem (P) involving the Lagrangian-type function $L_3 : X \times \mathbb{R}_+^m \times \mathbb{R}^r \rightarrow \mathbb{R}$ defined by

$$L_3(x,\mu,\vartheta) = \max_{1\leq i\leq p} \frac{f_i(x)}{q_i(x)} + \max_{1\leq i\leq p} \frac{\sum_{j=1}^{m}\mu_j g_j(x) + \sum_{k=1}^{r}\vartheta_k h_k(x)}{q_i(x)}.$$

We observe that, by Lemma 5, it follows that

$$L_3(x,\mu,\vartheta) = \max_{\lambda\in\Lambda} \frac{\sum_{i=1}^{p} \lambda_i f_i(x)}{\sum_{i=1}^{p} \lambda_i q_i(x)} + \max_{\lambda\in\Lambda} \frac{\sum_{j=1}^{m} \mu_j g_j(x) + \sum_{k=1}^{r} \vartheta_k h_k(x)}{\sum_{i=1}^{p} \lambda_i q_i(x)}.$$

The Lagrange-type function L_3 was first used by Xu [36] for formulating and proving saddle-point-type optimality and Lagrangian-type duality results for a minimax fractional programming problem with convex-concave ratios and convex constraints. However, Xu considered a generalized fractional programming problem with inequality constraints only. Now, in a natural way, we extend the definition of the Lagrangian L_3 for the considered generalized minimax fractional programming problem with both inequality and equality constraints.

Theorem 12. *Let $\overline{x} \in D$ be a normal optimal solution of the considered nonsmooth discrete minimax fractional programming problem (P) with the associated vector $\overline{\lambda}$ and $\overline{v}$ specified in Theorem 6. Further, assume that either one of the sets of hypotheses set forth in Theorem 8 is satisfied. Then*

$$L_3(\overline{x},\mu,\overline{\vartheta}) \leq L_3(\overline{x},\overline{\mu},\overline{\vartheta}) \leq L_3(x,\overline{\mu},\overline{\vartheta}) \text{ for all } x \in X \text{ and } \mu \in \mathbb{R}_+^m. \tag{38}$$

Proof. Since all assumptions of Theorem 8 are satisfied, therefore, the following inequalities

$$L_{1(\overline{v})}(\overline{x},w,\overline{\vartheta}) \leq L_{1(\overline{v})}(\overline{x},\overline{\mu},\overline{\vartheta}) \leq L_{1(\overline{v})}(x,\overline{\mu},\overline{\vartheta}) \tag{39}$$

hold for all $x \in X$ and $w \in \mathbb{R}_+^m$ with $\overline{v} = \varphi(\overline{x})$. Further, we have

$$\max_{1\leq i\leq p} [f_i(\overline{x}) - \overline{v}q_i(\overline{x})] = 0 \Leftrightarrow \overline{v} = \varphi(\overline{x}). \tag{40}$$

Hence, by (40) and the necessary optimality condition (5), it follows that

$$0 = \max\nolimits_{1\leq i\leq p}[f_i(\overline{x}) - \overline{v}q_i(\overline{x})] + \sum_{j=1}^{m} \overline{\mu}_j g_j(\overline{x}) + \sum_{k=1}^{r} \overline{\vartheta}_k h_k(\overline{x}) =$$
$$\max\nolimits_{1\leq i\leq p} \frac{f_i(\overline{x})}{q_i(\overline{x})} - \overline{v} + \max\nolimits_{1\leq i\leq p} \frac{\sum_{j=1}^{m} \overline{\mu}_j g_j(\overline{x}) + \sum_{k=1}^{r} \overline{\vartheta}_k h_k(\overline{x})}{q_i(\overline{x})} = L_3(\overline{x},\overline{\mu},\overline{\vartheta}) - \overline{v}, \tag{41}$$

where the last relation follows from the definition of the Lagrange function L_3. Next, letting $\omega = \max_{1\leq i\leq p}[1/q_i(\overline{x})]$,$\mu_j = w_j/\omega$, $j \in J$, and using the first inequality of (39) together with $\overline{x} \in D$, we get that the following inequalities

$$0 \geq \max\nolimits_{1\leq i\leq p}[f_i(\overline{x}) - \overline{v}q_i(\overline{x})] + \sum_{j=1}^{m} w_j g_j(\overline{x}) + \sum_{k=1}^{r} \overline{\vartheta}_k h_k(\overline{x}) \geq$$
$$\max\nolimits_{1\leq i\leq p} \frac{f_i(\overline{x})}{q_i(\overline{x})} - \overline{v} + \max\nolimits_{1\leq i\leq p} \frac{\sum_{j=1}^{m} \mu_j g_j(\overline{x}) + \sum_{k=1}^{r} \overline{\vartheta}_k h_k(\overline{x})}{q_i(\overline{x})} = L_3(\overline{x},\mu,\overline{\vartheta}) - \overline{v} \tag{42}$$

hold for any $\mu \in \mathbb{R}^m_+$. Combining (41) and (42), we obtain that the first inequality of (38) is fulfilled for any $\mu \in \mathbb{R}^m_+$.

Now, we prove the second inequality in (38). Let $x \in X$ be fixed and let $i(x) \in I$ be such that

$$f_{i(x)}(x) - \overline{v} q_{i(x)}(x) = \max_{1 \le i \le p} [f_i(x) - \overline{v} q_i(x)]. \tag{43}$$

Then, since $i(x)$ is fixed and $q_{i(x)}(x) > 0$, using the second inequality of (39), we get

$$\frac{\max_{1 \le i \le p}[f_i(x) - \overline{v} q_i(x)] + \sum_{j=1}^m \overline{\mu}_j g_j(x) + \sum_{k=1}^r \overline{\vartheta}_k h_k(x)}{q_{i(x)}(x)} \ge 0. \tag{44}$$

By (43), inequality (44) gives

$$\frac{f_{i(x)}(x)}{q_{i(x)}(x)} - \overline{v} + \frac{\sum_{j=1}^m \overline{\mu}_j g_j(x) + \sum_{k=1}^r \overline{\vartheta}_k h_k(x)}{q_{i(x)}(x)} \ge 0. \tag{45}$$

Note that

$$\begin{aligned} \max\nolimits_{1 \le i \le p} \frac{f_i(x)}{q_i(x)} - \overline{v} + \max\nolimits_{1 \le i \le p} \frac{\sum_{j=1}^m \overline{\mu}_j g_j(x) + \sum_{k=1}^r \overline{\vartheta}_k h_k(x)}{q_i(x)} \ge \\ \frac{f_{i(x)}(x)}{q_{i(x)}(x)} - \overline{v} + \frac{\sum_{j=1}^m \overline{\mu}_j g_j(x) + \sum_{k=1}^r \overline{\vartheta}_k h_k(x)}{q_{i(x)}(x)}. \end{aligned} \tag{46}$$

Hence, combining (45) and (46), we have

$$\max_{1 \le i \le p} \frac{f_i(x)}{q_i(x)} - \overline{v} + \max_{1 \le i \le p} \frac{\sum_{j=1}^m \overline{\mu}_j h_j(x) + \sum_{k=1}^r \overline{\vartheta}_k h_k(x)}{q_i(x)} \ge 0.$$

By the definition of the Lagrange function L_3, it follows that the inequality $L_3(x, \overline{\mu}, \overline{\vartheta}) \ge \overline{v}$ is satisfied for all $x \in X$. Then, by (41), the second inequality of (38) is satisfied for all $x \in X$. This completes the proof of the theorem. □

Theorem 13. *Let $\overline{x} \in D$ and, moreover, assume that there exists $\overline{\mu} \in \mathbb{R}^m_+$ and $\overline{\vartheta} \in \mathbb{R}^r_+$ such that the inequalities (38) are satisfied for all $x \in X$ and any $\mu \in \mathbb{R}^m_+$. Then $\overline{\mu}_j g_j(\overline{x}) = 0$, $j \in J$, and $\overline{x}$ is an optimal solution of the considered nonsmooth discrete minimax fractional programming problem (P).*

Proof. Proof of this theorem is similar to that of Theorem 9. □

In order to state our fourth saddle-point-type optimality result, we make use of the function $L_4 : X \times \Lambda \times \mathbb{R}^m_+ \times \mathbb{R}^r \to \mathbb{R}$ defined by

$$L_4(x, \lambda, \mu, \vartheta) = \frac{\sum_{i=1}^p \lambda_i f_i(x) + \sum_{j=1}^m \mu_j g_j(x) + \sum_{k=1}^r \vartheta_k h_k(x)}{\sum_{i=1}^p \lambda_i q_i(x)}.$$

The Lagrangian-type function given above was originally identified in [40] for a generalized fractional program containing convex n-set functions with the help of a Gordan-type transposition theorem.

Theorem 14. *Let $\overline{x}$ be a normal optimal solution of the considered nonsmooth discrete minimax fractional programming problem (P) with the associated vectors $\overline{\lambda}$, $\overline{\mu}$ and $\overline{\vartheta}$ specified in Theorem 6. Further, assume that either one of the following sets of hypotheses is satisfied:*

(a) *(i) for each $i \in I$, f_i is (Ψ, Φ, ρ_{f_i})-univex and $-q_i$ is (Ψ, Φ, ρ_{q_i})-univex at $\overline{x}$ on X;*

(ii) for each $j \in J$, g_j is (Ψ, Φ, ρ_{g_j})-univex at $\overline{x}$ on X;

(iii) for each $k \in K^+ = \{k \in K : \overline{\vartheta}_k > 0\}$, h_k is $(\Psi, \Phi, \rho^+_{h_k})$-univex at $\overline{x}$ on X;

(iv) for each $k \in K^- = \{k \in K : \overline{\vartheta}_k < 0\}$, $-h_k$ is $(\Psi, \Phi, \rho^-_{h_k})$-univex at $\overline{x}$ on X;

(v) the function Ψ is superlinear and $\Psi(a) \geq 0 \Rightarrow a \geq 0$;

(vi) $\sum_{i=1}^p \overline{\lambda}_i [M(\overline{x}, \overline{\lambda})\rho_{f_i} + N(\overline{x}, \overline{\lambda})\rho_{q_i}] +$
$M(\overline{x}, \overline{\lambda}) \left[\sum_{j=1}^m \overline{\mu}_j \rho_{g_j} + \sum_{k \in K^+} \overline{\vartheta}_k \rho^+_{h_k} - \sum_{k \in K^-} \overline{\vartheta}_k \rho^-_{h_k}\right] \geq 0.$

(b) *(i) the function*

$$z \to \Gamma_4(z, \overline{x}, \overline{\lambda}, \overline{\mu}, \overline{\vartheta}) = M(\overline{x}, \overline{\lambda}) \Big[\sum_{i=1}^p \overline{\lambda}_i f_i(z) + \sum_{j=1}^m \overline{\mu}_j g_j(z) + \sum_{k=1}^r \overline{\vartheta}_k h_k(x) \Big] - N(\overline{x}, \overline{\lambda}) \sum_{i=1}^p \overline{\lambda}_i q_i(z)$$

is $(\Psi_{\Gamma_4}, \Phi_{\Gamma_4}, \rho_{\Gamma_4})$-pseudounivex at $\overline{x}$ on X, where Ψ_{Γ_4} is an increasing function with $\Psi_{\Gamma_4}(a) \geq 0 \Rightarrow a \geq 0$ and $\rho_{\Gamma_4} \geq 0$.

Then, for all $x \in X$ and any $\mu \in \mathbb{R}^m_+$, it follows that

$$L_4(\overline{x}, \overline{\lambda}, \mu, \overline{\vartheta}) \leq L_4(\overline{x}, \overline{\lambda}, \overline{\mu}, \overline{\vartheta}) \leq L_4(x, \overline{\lambda}, \overline{\mu}, \overline{\vartheta}). \tag{47}$$

Proof. (a) By Theorem 6, there exist Lagrange multipliers $\overline{\lambda} \in \Lambda$, $\overline{\mu} \in \mathbb{R}^m_+$ and $\overline{\vartheta} \in \mathbb{R}^r$ such that (3)-(5) are satisfied. By (4), it follows that $\overline{v} = N(\overline{x}, \overline{\lambda})/M(\overline{x}, \overline{\lambda})$. If $\overline{v}$ is substituted into (3), then we have the following inclusion

$$0 \in M(\overline{x}, \overline{\lambda})\Big[\sum_{i=1}^p \overline{\lambda}_i \partial f_i(\overline{x}) + \sum_{j=1}^m \overline{\mu}_j \partial g_j(\overline{x}) + \sum_{k=1}^r \vartheta_k \partial h_k(\overline{x})\Big] \\ -N(\overline{x}, \overline{\lambda}) \sum_{i=1}^p \overline{\lambda}_i \partial q_i(\overline{x}).$$

Hence, it implies that there exist $\xi_i \in \partial f_i(\overline{x})$, $\zeta_i \in \partial(-q_i)(\overline{x})$, $i \in I$, $\varsigma_j \in \partial g_j(\overline{x})$, $j \in J$, and $\delta_k \in \partial h_k(\overline{x})$, $k \in K$, such that

$$M(\overline{x}, \overline{\lambda})\Big[\sum_{i=1}^p \overline{\lambda}_i \xi_i + \sum_{j=1}^m \overline{\mu}_j \varsigma_j + \sum_{k=1}^r \overline{\vartheta}_k \delta_k\Big] - N(\overline{x}, \overline{\lambda}) \sum_{i=1}^p \overline{\lambda}_i \zeta_i = 0. \tag{48}$$

Using hypotheses specified in (i)-(iv), by Definition 3, (13)-(17) hold. Since $\overline{\lambda} \geq 0$, $\overline{\mu} \geq 0$, $M(\overline{x}, \overline{\lambda}) > 0$, $N(\overline{x}, \overline{\lambda}) \geq 0$, and, by hypothesis (v), Ψ is a superlinear function, (13)-(17) yield

$$\Psi\Big(\sum_{i=1}^p \overline{\lambda}_i M(\overline{x}, \overline{\lambda})[f_i(x) - f_i(\overline{x})] + \sum_{i=1}^p \overline{\lambda}_i N(\overline{x}, \overline{\lambda})[-q_i(x) + q_i(\overline{x})] + \\ \sum_{j=1}^m \overline{\mu}_j M(\overline{x}, \overline{\lambda})[g_j(x) - g_j(\overline{x})] + \sum_{k=1}^r \overline{\vartheta}_k M(\overline{x}, \overline{\lambda}) \left[h_k(x) - h_k(\overline{x})\right] \Big) \geq \\ \sum_{i=1}^p \overline{\lambda}_i M(\overline{x}, \overline{\lambda}) \Phi(x, \overline{x}; (\xi_i, \rho_{f_i})) + N(\overline{x}, \overline{\lambda}) \sum_{i=1}^p \overline{\lambda}_i \Phi(x, \overline{x}; (-\zeta_i, \rho_{g_i})) + \\ M(\overline{x}, \overline{\lambda}) \Big[\sum_{j=1}^m \overline{\mu}_j \Phi(x, \overline{x}; (\varsigma_j, \rho_{h_j})) + \sum_{k \in K^+} \overline{\vartheta}_k \Phi(x, \overline{x}; (\delta_k, \rho^+_{h_k}) - \\ \sum_{k \in K^-} \overline{\vartheta}_k \Phi(x, \overline{x}; (-\delta_k, \rho^-_{h_k}) \Big]. \tag{49}$$

Let us define

$$\tau = \sum_{i=1}^p \overline{\lambda}_i [M(\overline{x}, \overline{\lambda}) + N(\overline{x}, \overline{\lambda})] + M(\overline{x}, \overline{\lambda}) \left[\sum_{j=1}^m \overline{\mu}_j + \sum_{k \in K^+} \overline{\vartheta}_k - \sum_{k \in K^-} \overline{\vartheta}_k \right], \tag{50}$$

$$\overline{\alpha}_i^M = \frac{M(\overline{x}, \overline{\lambda}) \overline{\lambda}_i}{\tau},\ \overline{\alpha}_i^N = \frac{N(\overline{x}, \overline{\lambda}) \overline{\lambda}_i}{\tau}, i \in I, \overline{\beta}_j = \frac{M(\overline{x}, \overline{\lambda}) \overline{\mu}_j}{\tau},\ j \in J, \\ \overline{\gamma}_k = \frac{M(\overline{x}, \overline{\lambda}) \overline{\vartheta}_k}{\tau}, k \in K^+, \overline{\gamma}_k = \frac{-M(\overline{x}, \overline{\lambda}) \overline{\vartheta}_k}{\tau}, k \in K^-. \tag{51}$$

Therefore, note that

$$\tau > 0, 0 \leq \overline{\alpha}_i^N \leq 1,\ 0 \leq \overline{\alpha}_i^M \leq 1\ \ 0 \leq \overline{\beta}_j \leq 1,\ 0 \leq \overline{\gamma}_k \leq 1, \tag{52}$$

and

$$\sum_{i=1}^{p}(\overline{\alpha}_i^M + \overline{\alpha}_i^N) + \sum_{j=1}^{m}\overline{\beta}_j + \sum_{k\in K^+}\overline{\gamma}_k - \sum_{k\in K^-}\overline{\gamma}_k = 1. \tag{53}$$

Now, dividing (49) by τ, using the necessary optimality conditions (4) and (5) together with the superlinearity of Ψ, we get

$$\Psi\Big(\sum_{i=1}^{p}[\overline{\alpha}_i^M f_i(x) - \overline{\alpha}_i^N g_i(x)] + \sum_{j=1}^{m}\overline{\beta}_j g_j(x) + \sum_{k\in K^+}\overline{\gamma}_k h_k(x) - \sum_{k\in K^-}\overline{\gamma}_k h_k(x)\Big) \geq \sum_{i=1}^{p}\overline{\alpha}_i^M \Phi(x,\overline{x};(\xi_i,\rho_{f_i})) + \sum_{i=1}^{p}\overline{\alpha}_i^N \Phi(x,\overline{x};(-\zeta_i,\rho_{q_i})) + \sum_{j=1}^{m}\overline{\beta}_j \Phi(x,\overline{x};(\varsigma_j,\rho_{g_j})) + \sum_{k\in K^+}\overline{\gamma}_k \Phi(x,\overline{x};(\delta_k,\rho_{h_k}^+)) + \sum_{k\in K^-}\overline{\gamma}_k \Phi(x,\overline{x};(-\delta_k,\rho_{h_k}^-)).$$

Using convexity of the function $\Phi(x,\overline{x};(\cdot,\cdot))$ together with (50)-(51), the above inequality yields

$$\Phi\Big(x,\overline{x};\Big(\sum_{i=1}^{p}(\overline{\alpha}_i^M \xi_i - \overline{\alpha}_i^N \zeta_i) + \sum_{j=1}^{m}\overline{\beta}_j \varsigma_j + \sum_{k\in K^+}\overline{\gamma}_k \delta_k - \sum_{k\in K^-}\overline{\gamma}_k \delta_k\ ,\ \sum_{i=1}^{p}[\overline{\alpha}_i^M \rho_{f_i} + \overline{\alpha}_i^N \rho_{q_i}] + \sum_{j=1}^{m}\overline{\beta}_j \rho_{g_j} + \sum_{k\in K^+}\overline{\gamma}_k \rho_{h_k}^+ + \sum_{k\in K^-}\overline{\gamma}_k \rho_{h_k}^-\Big)\Big) \leq \Psi\Big(\sum_{i=1}^{p}[\overline{\alpha}_i^M f_i(x) - \overline{\alpha}_i^N g_i(x)] + \sum_{j=1}^{m}\overline{\beta}_j g_j(x) + \sum_{k\in K^+}\overline{\gamma}_k h_k(x) - \sum_{k\in K^-}\overline{\gamma}_k h_k(x)\Big).$$

Since (52) and (53) are satisfied, using convexity of the function $\Phi(x,\overline{x};(\cdot,\cdot))$, the inequality above gives

$$\Phi\Big(x,\overline{x};\Big(\sum_{i=1}^{p}(\overline{\alpha}_i^M \xi_i - \overline{\alpha}_i^N \zeta_i) + \sum_{j=1}^{m}\overline{\beta}_j \varsigma_j + \sum_{k\in K^+}\overline{\gamma}_k \delta_k - \sum_{k\in K^-}\overline{\gamma}_k \delta_k\ ,\ \sum_{i=1}^{p}[\overline{\alpha}_i^M \rho_{f_i} + \overline{\alpha}_i^N \rho_{q_i}] + \sum_{j=1}^{m}\overline{\beta}_j \rho_{g_j} + \sum_{k\in K^+}\overline{\gamma}_k \rho_{h_k}^+ + \sum_{k\in K^-}\overline{\gamma}_k \rho_{h_k}^-\Big)\Big) \leq \Psi\Big(\sum_{i=1}^{p}[\overline{\alpha}_i^M f_i(x) - \overline{\alpha}_i^N g_i(x)] + \sum_{j=1}^{m}\overline{\beta}_j g_j(x) + \sum_{k\in K^+}\overline{\gamma}_k h_k(x) + \sum_{k\in K^-}\overline{\gamma}_k h_k(x)\Big).$$

Thus, (51) implies

$$\begin{aligned}
&\Phi\Big(x,\overline{x};\tfrac{1}{\tau}\Big(\textstyle\sum_{i=1}^{p}\overline{\lambda}_i\left(M(\overline{x},\overline{\lambda})\xi_i-N(\overline{x},\overline{\lambda})\zeta_i\right)+\\
&M(\overline{x},\overline{\lambda})\Big[\textstyle\sum_{j=1}^{m}\overline{\mu}_j\varsigma_j+\sum_{k=1}^{r}\overline{\vartheta}_k\delta_k\Big],\textstyle\sum_{i=1}^{p}\overline{\lambda}_i[M(\overline{x},\overline{\lambda})\rho_{f_i}+N(\overline{x},\overline{\lambda})\rho_{q_i}]\\
&+M(\overline{x},\overline{\lambda})\Big[\textstyle\sum_{j=1}^{m}\overline{\mu}_j\rho_{g_j}+\sum_{k\in K^+}\overline{\vartheta}_k\rho_{h_k}^{+}-\sum_{k\in K^-}\overline{\vartheta}_k\rho_{h_k}^{-}\Big]\Big)\Big)\leq \qquad (54)\\
&\Psi\Big(\tfrac{1}{\tau}\Big(\textstyle\sum_{i=1}^{p}\overline{\lambda}_i[M(\overline{x},\overline{\lambda})f_i(x)-N(\overline{x},\overline{\lambda})g_i(x)]+\\
&M(\overline{x},\overline{\lambda})\Big[\textstyle\sum_{j=1}^{m}\overline{\mu}_jg_j(x)+\sum_{k=1}^{r}\overline{\vartheta}_kh_k(x)\Big]\Big)\Big).
\end{aligned}$$

Since $\Phi(x,\overline{x};(0,a))\geq 0$ for all $a\in\mathbb{R}_+$, by (48) and hypothesis (vi), (54) yields

$$\begin{aligned}
&\Psi\Big(\tfrac{1}{\tau}\Big(\textstyle\sum_{i=1}^{p}\overline{\lambda}_i[M(\overline{x},\overline{\lambda})f_i(x)-N(\overline{x},\overline{\lambda})q_i(x)]+\\
&\qquad M(\overline{x},\overline{\lambda})\Big[\textstyle\sum_{j=1}^{m}\overline{\mu}_jg_j(x)+\sum_{k=1}^{r}\overline{\vartheta}_kh_k(x)\Big]\Big)\Big)\geq 0.
\end{aligned}$$

From $\Psi(a)\geq 0\Rightarrow a\geq 0$, the inequality above gives

$$\begin{aligned}
&\textstyle\sum_{i=1}^{p}\overline{\lambda}_i[M(\overline{x},\overline{\lambda})f_i(x)-N(\overline{x},\overline{\lambda})g_i(x)]+\\
&\qquad M(\overline{x},\overline{\lambda})\Big[\textstyle\sum_{j=1}^{m}\overline{\mu}_jg_j(x)+\sum_{k=1}^{r}\overline{\vartheta}_kh_k(x)\Big]\geq 0.
\end{aligned}\qquad (55)$$

Hence, (55) implies

$$\frac{\sum_{i=1}^{p}\overline{\lambda}_if_i(x)+\sum_{j=1}^{m}\overline{\mu}_jg_j(x)+\sum_{k=1}^{r}\overline{\vartheta}_kh_k(x)}{\sum_{i=1}^{p}\overline{\lambda}_iq_i(x)}\geq\frac{N(\overline{x},\overline{\lambda})}{M(\overline{x},\overline{\lambda})}.\qquad (56)$$

By definition of the Lagrange function L_4, it follows that

$$L_4(x,\overline{\lambda},\overline{\mu},\overline{\vartheta})\geq\frac{N(\overline{x},\overline{\lambda})}{M(\overline{x},\overline{\lambda})}=\frac{\sum_{i=1}^{p}\overline{\lambda}_if_i(\overline{x})}{\sum_{i=1}^{p}\overline{\lambda}_iq_i(\overline{x})}.\qquad (57)$$

By $\overline{x}\in D$ and the necessary optimality condition (5), (57) implies

$$L_4(x,\overline{\lambda},\overline{\mu},\overline{\vartheta})\geq\frac{N(\overline{x},\overline{\lambda})}{M(\overline{x},\overline{\lambda})}=\frac{\sum_{i=1}^{p}\overline{\lambda}_if_i(\overline{x})+\sum_{j=1}^{m}\overline{\mu}_jg_j(\overline{x})+\sum_{k=1}^{r}\overline{\vartheta}_kh_k(\overline{x})}{\sum_{i=1}^{p}\overline{\lambda}_iq_i(\overline{x})}.$$

By definition of the Lagrange function L_4, it follows that $L_4(x,\overline{\lambda},\overline{\mu})\geq L_4(\overline{x},\overline{\lambda},\overline{\mu})$ is satisfied for all $x\in X$, which is the second inequality of (47).

(b) Since (43) is satisfied, by Definition 4, $(\Psi_{\Gamma_4}, \Phi_{\Gamma_4}, \rho_{\Gamma_4})$-pseudounivexity implies that

$$\Psi\left(\Gamma_4(x, \overline{x}, \overline{\lambda}, \overline{\mu}, \overline{\vartheta}) - \Gamma_4(\overline{x}, \overline{x}, \overline{\lambda}, \overline{\mu}, \overline{\vartheta})\right) \geq 0.$$

Taking into account that $\Psi(a) \geq 0 \Rightarrow a \geq 0$, the inequality above yields

$$\Gamma_4(x, \overline{x}, \overline{\lambda}, \overline{\mu}, \overline{\vartheta}) \geq \Gamma_4(\overline{x}, \overline{x}, \overline{\lambda}, \overline{\mu}, \overline{\vartheta}).$$

But from (7) and (8) one can easily see that the right-hand side of this inequality is equal to zero and, therefore, it leads to (57). Hence the rest of the proof is identical to that of part (a). □

Theorem 15. *Let $\overline{x} \in D$. Further, assume that there exist $\overline{\lambda} \in \mathbb{R}^p_+$, $\overline{\mu} \in \mathbb{R}^m_+$ and $\overline{\vartheta} \in \mathbb{R}^r$ such that the inequalities (47) are satisfied for all $x \in X$ and any $\mu \in \mathbb{R}^m_+$. Then $\overline{\mu}_j g_j(\overline{x}) = 0$, $j \in J$ and, moreover, $\overline{x}$ is an optimal solution of the considered nonsmooth discrete minimax fractional programming problem (P).*

Proof. As in the proof of Theorem 9, by the first inequality of (47), it follows that

$$\sum_{j=1}^{m} \overline{\mu}_j g_j(\overline{x}) = 0. \tag{58}$$

In order to show that $\overline{x}$ is an optimal solution of problem (P), we make use of the second inequality in (47), that is, $L_4(\overline{x}, \overline{\lambda}, \overline{\mu}, \overline{\vartheta}) \leq L_4(x, \overline{\lambda}, \overline{\mu}, \overline{\vartheta})$ for all $x \in X$. Using $\overline{x} \in D$ together with (58), by the definition of the Lagrange function L_4, we get that the following inequality

$$\frac{\sum_{i=1}^{p} \overline{\lambda}_i f_i(\bar{x})}{\sum_{i=1}^{p} \overline{\lambda}_i q_i(\bar{x})} \leq \frac{\sum_{i=1}^{p} \overline{\lambda}_i f_i(x) + \sum_{j=1}^{m} \overline{\mu}_j g_j(x) + \sum_{k=1}^{r} \overline{\vartheta}_k h_k(x)}{\sum_{i=1}^{p} \overline{\lambda}_i q_i(x)} \tag{59}$$

holds for all $x \in X$. Therefore, it is also satisfied for all $x \in D$. Hence, for all $x \in D$, (59) yields

$$\frac{\sum_{i=1}^{p} \overline{\lambda}_i f_i(\bar{x})}{\sum_{i=1}^{p} \overline{\lambda}_i q_i(\bar{x})} \leq \frac{\sum_{i=1}^{p} \overline{\lambda}_i f_i(x)}{\sum_{i=1}^{p} \overline{\lambda}_i q_i(x)}. \tag{60}$$

If we choose $\overline{\lambda}_t = 1$ and $\overline{\lambda}_i = 0$ for all $i \in I \setminus \{t\}$ and repeat this process for $t = 1, 2, ..., p$, then (60) gives that the following inequalities

$$\frac{f_i(\bar{x})}{q_i(\bar{x})} \leq \frac{f_i(x)}{q_i(x)}, \, i = 1, ..., p \tag{61}$$

hold for all $x \in D$. Thus, (61) implies that the following inequality

$$\max_{1 \leq i \leq p} \frac{f_i(\bar{x})}{g_i(\bar{x})} \leq \max_{1 \leq i \leq p} \frac{f_i(x)}{g_i(x)}$$

holds for all $x \in D$. This means that $\overline{x} \in D$ is optimal in problem (P) and completes the proof of this theorem. □

Conclusion

In the paper, a new class of nonconvex nonsmooth minmax fractional programming problems with both inequality and equality constraints has been considered. Seven essentially equivalent Lagrangian-type functions have been defined for such an optimization problem. Under nondifferentiable (Ψ, Φ, ρ)-univexity and/or (Ψ, Φ, ρ)-pseudounivexity hypotheses, four sets of parametric and nonparametric saddle-point-type necessary and sufficient optimality conditions have been established for the considered nonsmooth minmax fractional programming problem involving locally Lipschitz functions. Thus, the equivalence between an optimal solution and a saddle point of seven Lagrange functions has been proved under assumptions that functions constituting the considered minmax fractional programming problem are nondifferentiable (Ψ, Φ, ρ)-univex and/or (Ψ, Φ, ρ)-pseudounivex.

Thus, to the best of our knowledge, the saddle point criteria presented in this paper are new in the area of nonconvex minmax fractional programming with locally Lipschitz functions and they are applicable for the largest class of nonconvex minmax fractional programming problems in comparison to those ones previously analyzed in the optimization literature.

References

[1] Ahmad, I. (2003). Optimality conditions and duality in fractional minimax programming involving generalized ρ-invexity. *International Journal of Statistics and Management System*, 19, 165-180.

[2] Ahmad, I., Gupta, S., Kailey, N., & Agarwal, R. P. (2011). Duality in nondifferentiable minimax fractional programming with B-(p, r)-invexity. *Journal of Inequalities and Applications*, 2011, article 75.

[3] Ahmad, I., & Husain, Z. (2006). Duality in nondifferentiable minimax fractional programming with generalized convexity. *Applied Mathematics and Computation*, 176, 545 -551.

[4] Ahmad, I., & Husain, Z. (2006). Optimality conditions and duality in nondifferentiable minimax fractional programming with generalized convexity. *Journal of Optimization Theory and Applications*, 129, 255-275.

[5] M.Al-roqi, A. (2015). Duality in minimax fractional programming problem involving nonsmooth generalized (F, α, ρ, d)-convexity. *Applied Mathematics & Information Sciences*, 9, 155-160.

[6] Antczak, T. (2008). Generalized fractional minimax programming with B-(p, r)-invexity. *Computers & Mathematics with Applications*, 56, 1505-1525.

[7] Antczak, T. (2011). Nonsmooth minimax programming under locally Lipschitz (Φ, ρ)-invexity. *Applied Mathematics and Computation*, 217, 9606-9624.

[8] Antczak, T. (2014). Parametric saddle point criteria in semi-infinite minimax fractional programming problems under (p, r)-invexity. *Numerical Functional Analysis and Optimization*, 35, 1511-1538.

[9] Antczak, T. & Stasiak, A. (2011). (Φ, ρ)-invexity in nonsmooth optimization. *Numerical Functional Analysis and Optimization*, 32, 1-25.

[10] Bătătorescu, A., Beldiman, M., Antonescu, I., & Ciumara, R. (2009). Optimality and duality for a class of nondifferentiable minimax fractional programming problems. *The Yugoslav Journal of Operations Research*, 19, 49-61.

[11] Bazaraa, M. S., Sherali, H. D., & Shetty, C. M. (1991). *Nonlinear programming: theory and algorithms*. New York: John Wiley and Sons.

[12] Bector, C. R., Chandra, S., & Bector, M. K. (2001). Generalized fractional programming duality: A parametric approach. *Journal of Optimization Theory and Applications*, 110, 611-619.

[13] Bector, C. R., Chandra, S., Gupta, S., & Suneja, S. K. (1994). Univex sets, functions and univex nonlinear programming. In: S. Komolosi, T.

Rapcsák, & S.Schaible, (Eds.), Generalized convexity. *Lecture Notes in Economics and Mathematical Systems*, vol.405, Berlin: Springer Verlag.

[14] Chandra, S., & Kumar, V. (1993). Equivalent Lagrangians for generalized fractional programming. *Opsearch*, 30, 193-203.

[15] Clarke, F. H., (1983). Optimization and nonsmooth analysis. *New York: A Wiley-Interscience Publication*, John Wiley&Sons, Inc.

[16] Dinkelbach, W. (1967). On nonlinear fractional programming. *Management Sciences*, 13, 492-498.

[17] Gramatovici, S. (2006). Duality for minimax fractional programming with generalized ρ-invexity. *Mathematical Reports*, 8, 151-166.

[18] Ho, S.-C. (2011). Sufficient conditions and duality theorems for nondifferentiable minimax fractional programming. *ISRN Mathematical Analysis*, Art. ID786978.

[19] Jayswal, A. (2008). Non-differentialble minimax fractional programming with generalized α-univexity. *Journal of Computational and Applied Mathematics*, 214, 121-135.

[20] Jayswal, A. (2011). Optimality and duality for nondifferentiable minimax fractional programming with generalized convexity. *ISRN Applied Mathematics*, 2011, Article ID 491941.

[21] Jayswal, A., Kummari, K., & Ahmad, I. (2015). Sufficiency and duality in minimax fractional programming with generalized (Φ, ρ)-invexity. *Mathematical Reports*, 17, 183-200.

[22] Jayswal, A., Prasad, A. K., & Kummari, K. (2013). On nondifferentiable minimax fractional programming involving higher order generalized convexity. *Filomat*, 27, 1497-1504.

[23] Lai, H.-Ch., & Ho, S.C. (2014). Mixed type duality on nonsmooth minimax fractional programming involving exponential (p, r)-invexity. *Numerical Functional Analysis and Optimization*, 35, 1-19.

[24] Lai, H.-Ch., & Lee, J. C. (2002). On duality theorems for a nondifferentiable minimax fractional programming. *Journal of Computational and Applied Mathematics*, 146, 115-126.

[25] Lai, H. C., & Liu, J. C., & Tanaka, K. (1999). Necessary and sufficient conditions for minimax fractional programming. *Journal of Mathematical Analysis and Applications*, 230, 311-328.

[26] Liang, Z., & Shi, Z. (2003). Optimality conditions and duality for a minimax fractional programming with generalized convexity. *Journal of Mathematical Analysis and Applications*, 277, 474-488.

[27] Liu, J. C. (1996). Optimality and duality for generalized fractional programming involving nonsmooth pseudoinvex functions. *Journal of Mathematical Analysis and Applications*, 202, 667-685.

[28] Liu, J. C., & Wu, C. S. (1998). On minimax fractional optimality conditions with (F, ρ)-convexity. *Journal of Mathematical Analysis and Applications*, 219, 36-51.

[29] Liu, X., & Yuan, D. (2014) Minimax fractional programming with nondifferentiable (G, β)-invexity. *Filomat*, 28, 2027-2035.

[30] Long, X.-J., Huang, N.-J., & Liu, Z.-B. (2008). Optimality conditions, duality and saddle points for nondifferentiable multiobjective fractional programs. *Journal of Industrial and Management Optimization*, 4, 287-298.

[31] Long, X. J., & Quan, J. (2011). Optimality conditions and duality for minimax fractional programming involving nonsmooth generalized univexity. *Numerical Algebra, Control and Optimization*, 1, 361-370.

[32] Luo, H.-Z., (2004). Incomplete Lagrange function and saddle point optimality criteria for a class of nondifferentiable generalized fractional programming. *Journal-Zhejiang University of Technology*, 32, 358-362.

[33] Mangasarian, O. L. (1969). *Nonlinear programming*. New York: McGraw-Hill.

[34] Chandra Sekhara, O., Reddy, C. M., & Raja Sekhara, R.P. (2009). Saddle-point optimality criteria and duality results for minimax multi-objective fractional programming problem. *Journal of Pure and Applied Physics*, 21, 587-601.

[35] Stancu-Minasian, I., Stancu, A. M., & Jayswal, A. (2016). Minimax fractional programming problem with (p, r)-ρ-(η, θ)-invex functions. *Annals of the University of Craiova, Mathematics and Computer Science Series*, 43, 94-107.

[36] Xu, Z. K. (1988). Saddle-point type optimality criteria for generalized fractional programming. *Journal of Optimization, Theory and Applications*, 57, 189-196.

[37] Yang, X. M., Yang, X. Q., & Teo, K. L. (2004). Duality and saddle-point type optimality for generalized nonlinear fractional programming. *Journal of Mathematical Analysis and Applications*, 289, 100-109.

[38] Yuan, D. H., Liu, X. L., Chinchuluun, A., & Pardalos, P. M. (2006). Nondifferentiable minimax fractional programming problems with (C, α, ρ, d)-convexity. *Journal of Optimization, Theory and Applications*, 129, 185-199.

[39] Zalmai, G. J. (1986). Optimality conditions for a class of nondifferentiable minmax programming problems. *Optimization*, 17, 453-465.

[40] Zalmai, G. J. (1990). Duality for generalized fractional programs involving n-set functions. *Journal of Mathematical Analysis and Applications*, 149, 339-350.

[41] Zalmai, G. J. (1995). Optimality conditions and duality models for generalized fractional programming problems containing locally subdifferentiable and ρ-convex functions. *Optimization*, 32, 95-124.

[42] Zalmai, G. J. (2006). Saddle points and Lagrangian-type duality for discrete minmax fractional subset programming problems with generalized convex functions. *Journal of Mathematical Analysis and Applications*, 313, 484-503.

[43] Zalmai, G. J., & Zhang, Q. (2012). Optimality conditions and duality in minmax fractional programming, part I: Necessary and sufficient optimality conditions. *Journal of Advances Mathematical Studies*, 2, 107-137.

In: Advances in Mathematics Research
Editor: Albert R. Baswell

ISBN: 978-1-53612-512-2

Chapter 4

BATTERY CHARGE AND DISCHARGE BEHAVIOR PREDICTION USING ELECTRICAL MATHEMATICAL MODELS

Marcia de F. B. Binelo, Leonardo B. Motyczka, Airam T. Z. R. Sausen, Paulo S. Sausen* and Manuel O. Binelo
Postgraduate Program in Mathematical Modeling,
Regional University of Northwestern
Rio Grande do Sul State (UNIJUÍ),
Ijuí - RS - Brazil

Abstract

Battery behavior modeling, under different use conditions, can be relatively complex due to the nonlinear nature of the charge and discharge processes. Understanding these dynamics by leveraging mathematical models, favors the development of more efficient batteries and also provides tools for software developers to better manage device resources. A review of electrical mathematical models used in the prediction of battery charge and discharge behavior is presented in this chapter. The class of electrical models has been used in various battery modeling applications, including mobile devices and electrical vehicles. The scientific

*Corresponding Author Email: sausen@unijui.edu.br.

investigation of such models is motivated by their capacity to provide important electrical information such as current, voltage, state of charge, and also some nonlinear aspects of the problem while keeping a relatively low complexity. Six subclasses of electrical models (Simple models, Thévenin-based models, Impedance-based models, Runtime-based models, Combined models and Generic models) along with a discussion of the main characteristics of each. This will demonstrate the evolution of electrical models through successive modification and combination, resulting in varying levels of accuracy and complexity.

Keywords: mathematical modeling, battery models, electrical models

1. Introduction

Rechargeable batteries are a vital component for many electrical systems, such as mobile devices, and Electrical Vehicles (EVs) [46]. The rapid pace of the mobile market is driven by the increasing desire for portability, usability, and mobility. The growing adoption of EVs is an irreversible trend due to the environmental problems caused by transport means powered by fossil fuels [54]. In order to ensure the satisfactory performance of these systems, it is necessary to use batteries able to provide the energy capacity demanded by such applications, taking into account the operational, economic and environmental aspects.

The battery behavior modeling, under various use conditions, can be relatively complex due to the nonlinearities present in the charge and discharge processes [40, 46]. The understanding of these dynamics by the use of mathematical models can favor the development of more efficient batteries, and also provide tools for software developers to better manage device resources. Two fundamental characteristics that must be considered are the high accuracy, and low complexity that are expected from the models applied in this context, in special, those used for battery lifetime prediction [40, 46].

Some aspects are desired for the development of accurate and easily applicable mathematical models. An important factor is the correct determination of the state of charge (*SOC*), defined as the battery capacity at a given moment in relation to the total capacity, expressed as a percentage [53]. Two nonlinear effects that occur during the discharge process that need to be taken into account are the recovery effect and the rate capacity effect [40]. Also, the temperature affects the open circuit voltage (*Voc*), the battery capacity and the battery internal resistance [54]. Another important effect is the battery capacity fading, that impacts the battery remaining usable capacity [15].

Several mathematical battery models have been developed with varying levels of accuracy and complexity, and are divided into the following classes: Electrochemical [14,48], Stochastic [9,37], Analytical [3,40], Electrical [8,19], and System Identification [44]. The class of Electrical models has been used in different battery modeling applications, including mobile devices and EVs [17,32,43]. These models combine different electric components such as voltage sources, resistors, capacitors, and nonlinear components, such as, the Warburg impedance [54]. The scientific investigation of said models is motivated by their capability of providing important electrical information like current, voltage and *SOC*, and also the fact that they can incorporate some nonlinear aspects of the problem while maintaining a relatively low complexity [19].

The Electrical model class can be further divided into subclasses. Some authors [8, 29] divide these models into Thévenin-based, Impedance-based and Runtime-based models. The Thévenin-based models provide the battery transient responses using resistor and capacitor networks (*RC*) [31]. Some Thévenin-based models are presented in [2,18,36]. The Impedance-based models are developed using the Electrochemical Impedance Spectroscopy (EIS) method, and can provide a good description of the battery's internal behavior [30]. The Runtime-based models are able to simulate the battery lifetime and voltage during the discharge processes [8]. In [35], three more classes are mentioned, the Simple, Combined, and Generic models. Simple models use only a voltage source and an internal resistance [17]. Combined models are able to simulate the battery lifetime, steady-state, and transient responses [8,54]. The Generic models can be used to model batteries of different technologies under different charge and discharge conditions [50].

In this chapter, a review of electrical models used in the prediction of battery charge and discharge behavior is presented. These models have an important role to drive innovation for battery design and operation management. Six subclasses of electrical models (Simple models, Thévenin-based models, Impedance-based models, Runtime-based models, Combined models and Generic models) along with a discussion of the main characteristics of each. This will demonstrate the evolution of electrical models through successive modification and combination, resulting in varying levels of accuracy and complexity.

2. Battery Electrical Models

In this section, the main characteristics of each electrical model's subclasses are present as well as the application of such models in predicting of battery charge and discharge behavior.

2.1. Simple Models

The equivalent circuit of a simple battery model is generally composed of an internal resistance R_0 and an ideal voltage source, usually represented by an open circuit voltage *Voc* [7]. One example of this subclass is the Simple model, described in [7] and shown in Figure 1, also known as Rint model [17] and Resistance model [46].

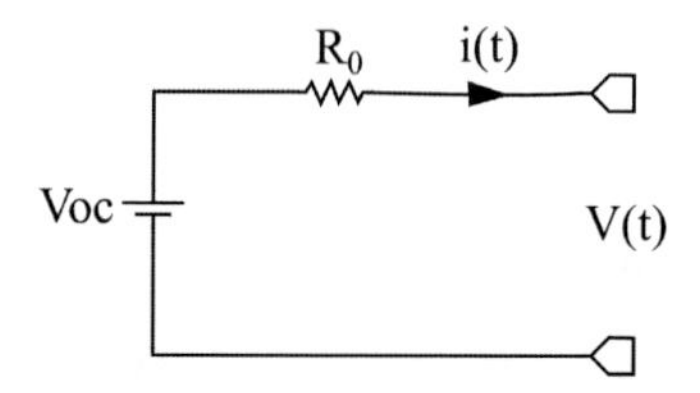

Figure 1. Simple model equivalent circuit.

By applying the Kirchhoff circuit law, the output voltage $V(t)$ is

$$V(t) = Voc - R_0 i(t), \tag{1}$$

where: $i(t)$ is the electric current drawn from the battery [17].

This model does not take into account the *SOC* and R_0 is assumed to be constant. Due to these assumptions, this model is only applicable in simulations where the energy extracted from the battery is considered to be limitless [7].

The accuracy of the Simple model can be improved if more factors are taken into account such as the dependency of R_0 and *Voc* on *SOC* [10, 24], and the inclusion of new components, such as a polarization constant, in order to adjust the battery parameters during the charge or discharge process [7, 33]. The performance of this model can also be improved with the addition of more components to the circuit, such as resistor-capacitor networks (*RC*), modeling the transient behavior of the battery [46]. These modifications characterize the Thévenin-based model subclass described below.

2.2. Thévenin-Based Models

Generally, these models present an important characteristic: the capability to capture two time dependent effects, the depletion effect and the recovery effect [46]. The charge depletion effect is observed during the first moments of battery discharge. Due to an initial high concentration of chemicals near the cathode and anode, the initial voltage drop is steep, and as these chemicals are consumed by the electrochemical reactions, the voltage drop becomes more gradual [13, 46]. The charge recovery effect occurs during time intervals when the discharge current is zero or severely reduced, allowing the reorganization of electrons in the electrolyte, which results in an effective increase in battery capacity [27]. This behavior can be observed in a battery discharge pulse, as shown in Figure 2.

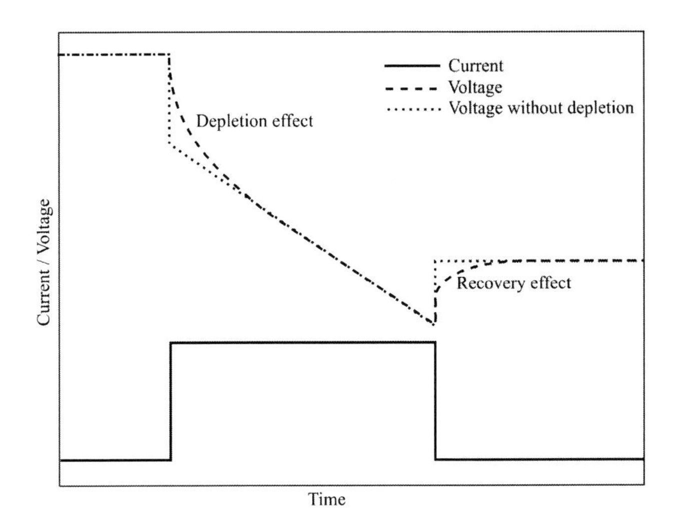

Figure 2. Charge depletion and charge recovery effects.

Six models based on Thévenin are presented in this subsection: Thévenin, DP, Linear Electrical, Resistive Thévenin, Reactive and Modified.

2.2.1. Thévenin Model

The Thévenin model, shown in Figure 3, is an extension of the Simple model with the addition of an *RC* network that models the transient response during the battery charge and discharge processes.

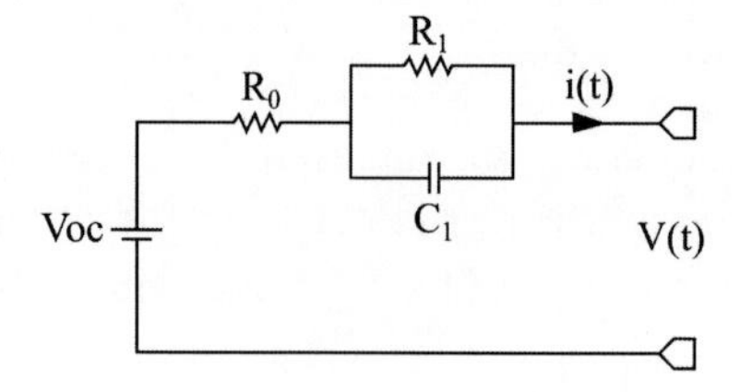

Figure 3. Thévenin model equivalent circuit.

The electrical behavior of this model can be expressed by

$$V(t) = Voc - R_0 i(t) - V_1, \tag{2}$$

where: V_1 is the voltage across the capacitor C_1, which is obtained by the Ordinary Differential Equation (ODE)

$$\frac{dV_1}{dt} = \frac{i(t)}{C_1} - \frac{V_1}{R_1 C_1}, \tag{3}$$

where: R_1 is the polarization resistance and C_1 is the corresponding capacitance [31]. Therefore, solving the equation (3) we have

$$V_1 = R_1 i(t)(1 - e^{-\frac{t}{R_1 C_1}}). \tag{4}$$

The main limitation of this model is that all parameters are constants [35, 43]. In actual charge and discharge processes, these values would vary with the battery work conditions and also with the battery usage history [43], since these parameters are related to *SOC*, load capacity, discharge rate and temperature [35].

Some studies were done to evaluate the ideal number of *RC* networks necessary for the model to achieve the adequate accuracy [20, 55]. A comparison among different electrical models applied to lithium ion battery modeling is presented in [20]. The results show that a model with one parallel *RC* network can describe the battery behavior with sufficient precision. However, [55] demonstrates that two parallel *RC* networks result in a good trade-off between high accuracy and low complexity, two important aspects necessary for the efficient application of a model.

2.2.2. DP Model

The DP model, introduced by [18], is an extension of the Thévenin [17] model, with the addition of an *RC* network, as can be seen in Figure 4. The new *RC* network is used to better describe the transient characteristics of the battery discharge [17].

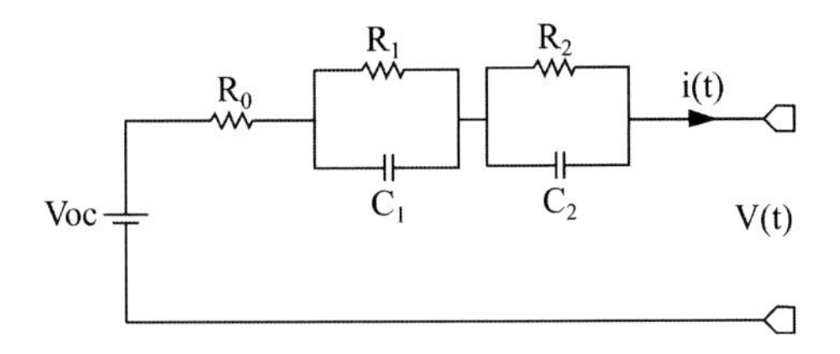

Figure 4. DP model equivalent circuit.

The output voltage $V(t)$ is

$$V(t) = Voc - R_0 i(t) - V_1 - V_2, \tag{5}$$

where: V_1 is obtained from the Thévenin model, and V_2 is the voltage across the capacitor C_2, given by

$$\frac{dV_2}{dt} = \frac{i(t)}{C_2} - \frac{V_2}{R_2 C_2}, \tag{6}$$

where: R_2 is the concentration polarization resistance and C_2 is the equivalent capacitance [18]. By solving the ODE presented in equation (6), we have

$$V_2 = R_2 i(t)(1 - e^{-\frac{t}{R_2 C_2}}). \tag{7}$$

In [17], seven battery models are compared, and the results show that the voltage relaxation effect, associated with the recovery effect, cannot be ignored. The Thévenin model and the DP model presented the best results when compared to the five other models (i.e., Shepherd model, Unnewehr Universal model, Nernst model, Combined model and Rint model), with the DP model presenting the most accurate results due to its more detailed simulation of the voltage relaxation.

2.2.3. Linear Electrical Model

Another model based on the Thévenin model is the Linear Electrical model [2], shown in Figure 5. This model takes into account the battery behavior in respect to the self-discharge and overvoltage processes only with linear components [35].

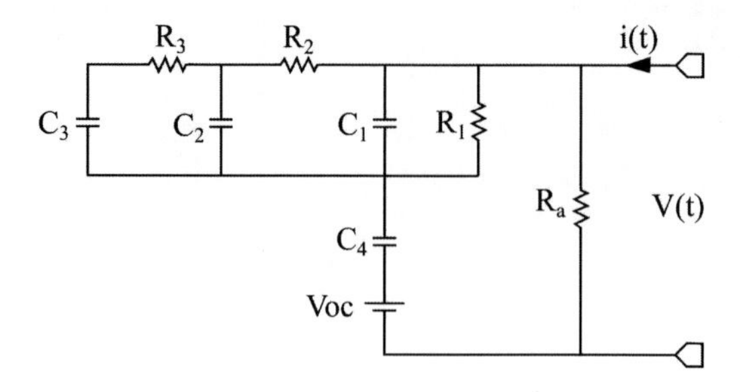

Figure 5. Linear Electrical model equivalent circuit.

The battery self-discharge is modeled by the R_a resistor in parallel to the charge storage capacitor C_4 [35]. The overvoltage network N_m is composed of 3 resistances (R_1, R_2, and R_3) and 3 capacitors (C_1, C_2, and C_3), and is calculated as

$$N_m = \sum_{i=1}^{3} R_i i(t) \left[1 - exp(-t/\tau_i)\right], \quad (8)$$

where: R_i are the equivalent resistances and τ_i are the time constants according to the RC networks. These parameters are set according to the battery electrical response obtained by experimental means [2].

The output voltage $V(t)$ is given by

$$V(t) = Voc + N_m. \quad (9)$$

Although this model is more accurate than those previously cited (Thévenin and DP models), it does not take into account the temperature, and uses different sets of parameter values to model the battery at different charge states. Therefore, a continuous simulation of the battery charge or discharge process is not viable [45]. However, this model can be used for the analysis the stationary and transient states of electrical systems powered by batteries [35].

2.2.4. Resistive Thévenin Model

The Resistive Thévenin model [47] is composed of an ideal voltage source connected to two internal resistances R_1 and R_2, respectively related to the battery charge and discharge processes. Diodes are employed to differentiate the resistances for charge or discharge cycles, assuming that both processes are not simultaneous. The model equivalent circuit is shown in Figure 6.

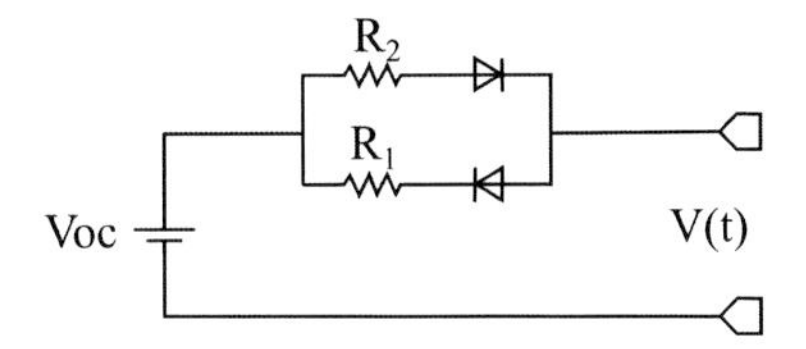

Figure 6. Resistive Thévenin model equivalent circuit.

Since there are no capacitors, the battery electrical behavior is based on the Simple model, described by equation (1). The dependency on *SOC* is not taken into account, and all parameters are linear. Therefore, it is not advisable to use the Resistive Thévenin model for electric/hybrid vehicles battery modeling due to the dynamic nature of the processes involved [35].

2.2.5. Reactive Model

A model similar to the Resistive Thévenin is presented in [36], and is called the Reactive model. The equivalent circuit, shown in Figure 7, has an internal resistance R_0 and a capacitance C_1 added, beyond the components already present in the previous model [35].

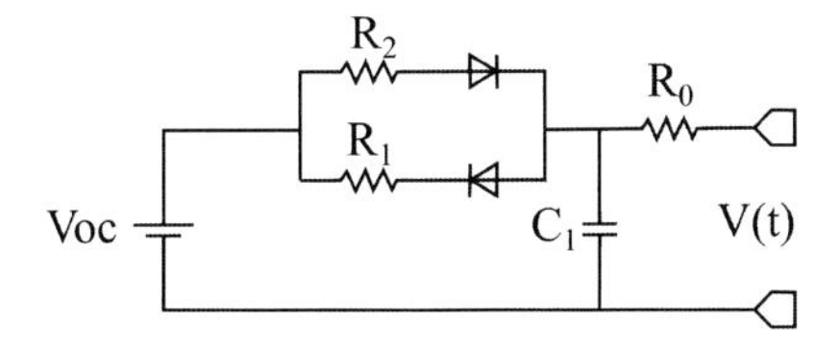

Figure 7. Reactive model equivalent circuit.

The electric dynamics of the charge and discharge cycles are modeled by the ODEs (10) and (11) [36], respectively,

$$\frac{dV_1}{dt} = -V_1 \frac{1}{R_1 C_1} + V_{oc} \frac{1}{R_1 C_1} - i(t) \frac{1}{C_1}, \quad V_1 > V_{oc}, \tag{10}$$

$$\frac{dV_1}{dt} = -V_1 \frac{1}{R_2 C_1} + V_{oc} \frac{1}{R_2 C_1} - i(t) \frac{1}{C_1}, \quad V_1 \leq V_{oc}, \tag{11}$$

where: V_1 is the voltage on capacitor C_1.

The current $i(t)$ can be obtained by

$$i(t) = \frac{V_1 - V_{oc}}{R_0}. \tag{12}$$

Therefore, the output voltage $V(t)$ is given by

$$V(t) = V_{oc} - R_0 i(t) - V_1. \tag{13}$$

It is important to note that the SOC can be calculated from the Voc when the following conditions are present: $R_0 = 0$ and $V(t) = V_1$. Both will exponentially converge to V_0 with a time constant defined by R_2. This model can be successfully applied on simulations of electrical and hybrid vehicle batteries.

2.2.6. Modified Model

The Modified model [6], presented in Figure 8, is a modification of the Reactive model, where the R_{co} overcharge and R_{do} over discharge resistances were added. When the battery is overcharged or over discharged, the internal resistance can significantly increase due to the electrolyte diffusion [7,35]. Also, the resistance R_0 was removed from the circuit and a self-discharge resistance R_a was added. In this way, the model takes into account the nonlinear characteristics of the charge and discharge process and the dependence of the model components on SOC [7].

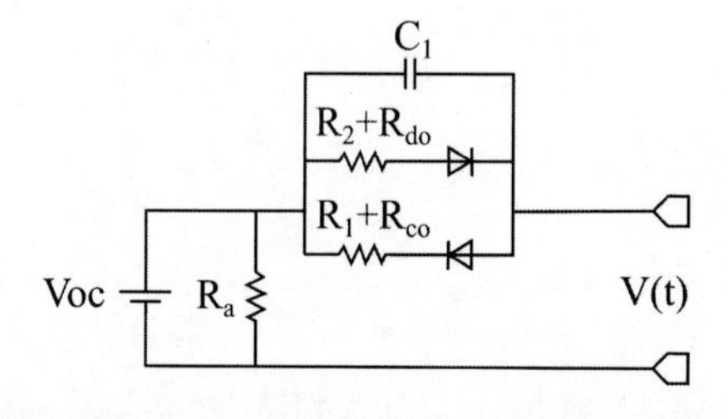

Figure 8. Modified model equivalent circuit.

An extension of this model can be found in [45], where two new internal resistances were added to better describe the battery behavior. The new model elements are a function of *SOC* and *Voc*. The diodes differentiate internal and overvoltage resistances. This model was proposed considering the analysis of plots obtained from experimental data and specifications from lead-acid battery manufacturers [35].

2.3. Impedance-Based Models

This subclass of electrical models is based on the Electrochemical Impedance Spectroscopy (EIS) technique, which consists of applying small-amplitude sinusoidal signals to the system and then measuring the response for different frequencies. This response is interpreted to obtain the battery's complex impedance, with the real part representing the resistance, and the imaginary part the reactance [35,46]. The frequency properties of the battery can be obtained by repeating this process for a given frequency interval [22,52].

Impedance-based models are considered one of the most accurate methods for the modeling of electrochemical systems, including batteries of different technologies [5, 52]. Using an electrode as reference, it is possible to measure and model the positive and negative electrode spectra separately. This can ensure better results, since the parameters cannot be trusted when jointly estimated [21]. Some tests, based on mathematical studies, are proposed in [11] to validate the measured impedance data, characterizing the system properties in time and frequency domains. The usual method to estimate the EIS model parameters is the use of nonlinar least squares [46], but other alternative methodologies, such as metaheuristics, have also been used [51].

However, some studies have shown that the EIS technique is too complex to be used in real time applications, since the EIS spectrum depends on *SOC*, current, temperature, and state-of-health (SOH) [46]. For some authors [26,39,51], this is considered a positive aspect of the method because these effects can then be included in the developed models. With the use of impedance models, a good description of the battery internal dynamics is expected, including the charge transference reactions on the electrode/electrolyte interface, the ion diffusion in the electrode, the double layer effects and the rise in the resistance and capacitance of the anode isolation film [30].

In this subsection, two impedance-based models are presented, the Randles and the Temporal models.

2.3.1. Randles Model

One of the first impedance-based models was proposed by Randles, in 1947 [42]. This model, shown in Figure 9, is composed of a resistance R_0, an *RC* network composed of the charge transfer resistance R_1 and the corresponding capacitor C_1; and the Warburg impedance Z_W representing the diffusion dynamic [29].

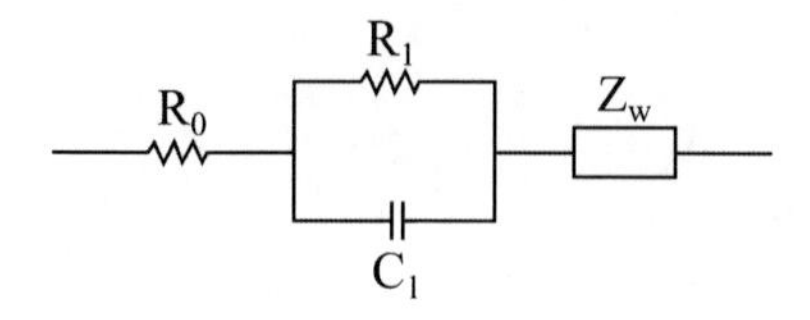

Figure 9. Randles model equivalent circuit.

The Warburg impedance Z_W can be obtained as

$$Z_W = \frac{A_W}{\sqrt{\omega}}(1-j), \tag{14}$$

where: A_W is the Warbung coefficient, ω is the angular frequency and j is the imaginary number $\sqrt{-1}$ [46]. This model has been used as the basis for the development of new models, where different expressions are used for the impedance definition [28,32,52].

An analysis of the Randles circuit using Laplace transform is done in [12], where the main equations describing the circuit's transient behavior were obtained. The authors proposed a time-frequency analysis of a pulsed excitation signal, so that the transient current through the capacitor is zero when the pulse is zero. In this way, it is possible to estimate the resistive component of an electrochemical cell.

The representation of the generalized Randles model is shown in Figure 10, and its identifiability analysis is done in [1], in order to verify if its parameters can be estimated using the input and output signals. In this model, the number of *RC* networks depends on how many pairs are necessary for the response to be adjusted to the device impedance spectrum in the desired frequency band.

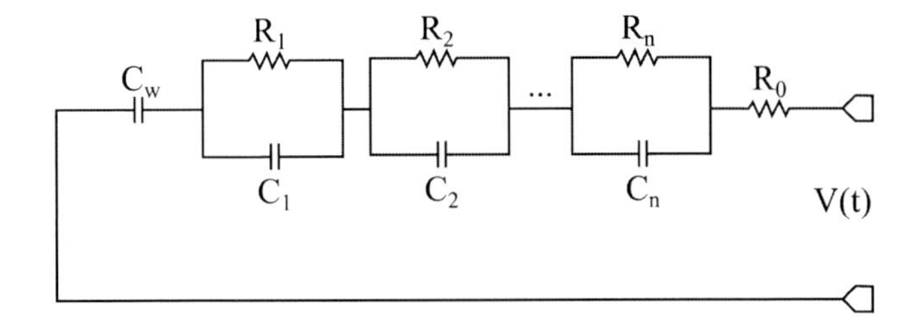

Figure 10. Generalized Randles model equivalent circuit.

2.3.2. Temporal Model

The Temporal model was developed by [29] for lithium ion batteries, and is an extension of the Randles model, as can be seen in Figure 11.

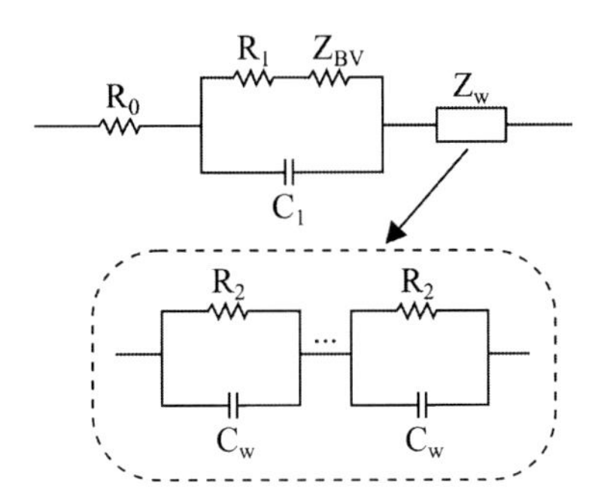

Figure 11. Temporal model equivalent circuit.

During a battery's continuous discharge operation, a thermodynamic effect caused by the active mass reduction in the porous electrode can be observed. This reduction can affect the electrode electrochemical reaction kinetics, and consequently, compromise the over-potential [29]. This effect can be incorporated to the model by the inclusion of the Butler-Volmer term, represented by the Z_{BV} impedance, given by [47]

$$Z_{BV} = k\frac{Q_m}{Q_m - \int i(t)dt}, \tag{15}$$

where: Q_m is the maximum amount of Li+ ions electrically active in the battery, and k is the electrode constant reaction rate.

To obtain the SOC information in real time, a temporal approximation is used, consisting of an RC circuit representing the Warburg element, defined in the frequency domain. Applying the inverse Laplace transform to the Warbung impedance, we have the temporal expression [26]

$$Z_W = \sum_{n=1}^{\infty} \frac{1}{C_W} exp \frac{-t}{R_n C_W}, \tag{16}$$

and

$$C_W = \frac{k_1}{2k_2^2}, \tag{17}$$

$$R_n = \frac{8k_1}{(2n-1)^2 \pi^2}, \tag{18}$$

where: k_1 and k_2 are obtained from the EIS, and n is the number of RC pairs necessary to represent the Warburg element.

The fitting of the circuit to the battery discharge curve was done using the Simulated Annealing (SA) and Levenberge-Marquart fitting Algorithm (LMA) methods. Different values of n were tried in order to find the correct amount of RC pairs, and also for the parameters estimation [29]. This model is able to provide, in real time, the battery maximum charge capacity after each discharge cycle, without the need to fully discharge it, avoiding its useful lifetime reduction.

2.4. Runtime-Based Models

Runtime-based models are generally composed of three different circuits, an example is shown in Figure 12. The first circuit is composed of a resistance R_1, a capacitor C_1 and a voltage source V_{c-rate}, and it provides the battery transient characteristics, resulting in the discharge rate V_{rate}. The second circuit represents the SOC dependence on the discharge rate, and is composed of a self-discharge resistance R_a, a capacitor C_2 representing the battery total stored charge, a controlled current source $i(t)$ and a voltage source V_{lost}. The V_{lost} is a function of the V_{rate}. The third circuit has a voltage source Voc representing the SOC and a resistance R_0, resulting in the voltage at the battery terminals [35].

With this model it is possible to simulate the battery lifetime and voltage during a direct current discharge process, but it might not present good results when simulating variable current discharges due to its low precision modeling

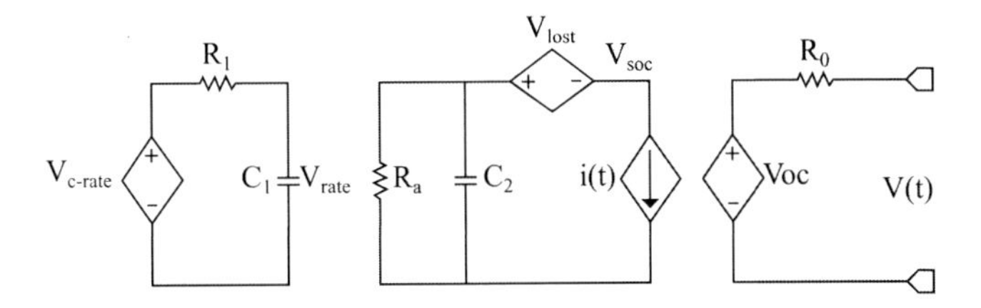

Figure 12. Runtime-Based model equivalent circuits.

of the battery transient characteristics [8]. In this model, the voltage depends on *SOC* and the capacity depends on the discharge rate. Even considering its limitations, this model has been successfully applied in the mathematical modeling of electrical and hybrid vehicles [35].

2.5. Combined Models

The combination of different electrical models makes it possible to take advantage of the positive attributes of each model, such as the accurate estimation of the *SOC*, transient response and temperature effects, and the correct prediction of the battery runtime [8, 25]. One of the main combined models developed is the Chen and Rincón-Mora model [8] next described. Two other models derived from this model are also presented in this subsection, the Kroeze and Krein model and the Erdinc model.

2.5.1. Chen and Rincón-Mora Model

The Chen and Rincón-Mora model [8] is an electrical model capable of predicting battery runtime and I-V characteristics, and is presented in Figure 13. This model combines the runtime-based models (circuit on the left) and the Thévenin-based models (circuit on the right). The circuit on the left is composed of R_a, C_3 and $i(t)$ and models the capacity, the *SOC* and the battery lifetime. The circuit on the right models the transient response using a resistance R_0, two parallel *RC* networks, and a controlled voltage source $Voc(V_{SOC})$. This model has good accuracy and has been applied on the modeling of batteries of different technologies [8].

Two time constants, one of short and other of long duration, are used to model the transient response. They are defined by the interval $t_0 < t < t_r$. During the time interval $t_0 < t < t_d$, the battery is discharged with a constant current

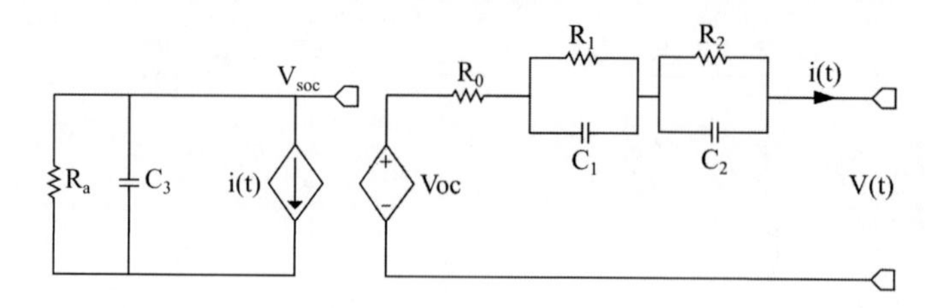

Figure 13. Chen and Rincón-Mora model equivalent circuits.

$(i(t) > 0)$, and during the time interval $t_d < t < t_r$, the battery is in a rest state $(i(t) = 0)$. The short and long transient responses are represented by the R_1, C_1, R_2 e C_2 resistances and capacitors. When in steady-state, C_1 and C_2 perform the open circuit role, offering a high resistance to the direct current, whereas in transient state, this capacitors behave as a short circuit until fully charged [8].

The battery usable capacity is found from the relation among C_3, R_a, R_0, R_1 and R_2 components. The value of C_3 is given by

$$C_3 = 3600Capacity f_1[n_{cycle}] f_2[T(t)], \tag{19}$$

where: *Capacity* is the nominal capacity, $f_1[n_{cycle}]$ is the number of cycles correction factor and $f_2[T(t)]$ is the temperature correction factor [8].

The output voltage $V(t)$ is expressed as

$$V(t) = Voc(V_{SOC}) - R_0 i(t) - V_{Transient}(t), \tag{20}$$

where: $V_{Transient}(t)$ is the transient voltage calculated from

$$V_{Transient}(t) = V_1 - V_2, \tag{21}$$

where: V_1 is the short duration transient voltage, and V_2 is the long duration transient voltage, both given by the equations (22) and (23), respectively,

$$V_1 = \begin{cases} R_1 i(t)\left[1 - e^{-\frac{(t-t_0)}{\tau_1}}\right], & t_0 < t < t_d, \\ V_1(t_d)e^{-\frac{(t-t_d)}{\tau_1}}, & t_d < t < t_r, \end{cases} \tag{22}$$

$$V_2 = \begin{cases} R_2 i(t)\left[1 - e^{-\frac{(t-t_0)}{\tau_2}}\right], & t_0 < t < t_d, \\ V_2(t_d)e^{-\frac{(t-t_d)}{\tau_2}}, & t_d < t < t_r, \end{cases} \tag{23}$$

where: $V_1(t_d)$ is the short duration transient voltage at final discharge time, $\tau_1 = R_1C_1$, $V_2(t_d)$ is the long duration transient voltage at the final discharge time, and $\tau_2 = R_2C_2$ [23].

The elements Voc and R_0 from equation (20), and also R_1, C_1, R_2 and C_2, that model the transient voltage response, are functions of SOC, as described by the following equations

$$Voc(SOC) = a_0e^{-a_1SOC} + a_2 + a_3SOC - a_4SOC^2 + a_5SOC^3, \tag{24}$$

$$R_0(SOC) = b_0e^{-b_1SOC} + b_2, \tag{25}$$

$$\begin{cases} R_1(SOC) = c_0e^{-c_1SOC} + c_2, \\ C_1(SOC) = d_0e^{-d_1SOC} + d_2, \\ R_2(SOC) = e_0e^{-e_1SOC} + e_2, \\ C_2(SOC) = f_0e^{-f_1SOC} + f_2. \end{cases} \tag{26}$$

The methodology usually adopted in the literature for the estimation of the 21 parameters of equations (24), (25) and (26) is based on curve fitting techniques, using 4 pulsed battery discharge curves [8].

2.5.2. Kroeze and Krein Model

The Kroeze and Krein model [25] is an extension of the Chen and Rincón-Mora model [8] with the inclusion of a resistance and an *RC* network, as can be seen in Figure 14. The resistance R_f was added to the left circuit to model the battery capacity correction factor, in this way modeling the capacity dependence on current. The *RC* network added to the circuit on the right, was used to model the transient response with duration inferior to 1 second. Therefore, this model uses three time constants for seconds, minutes and hours.

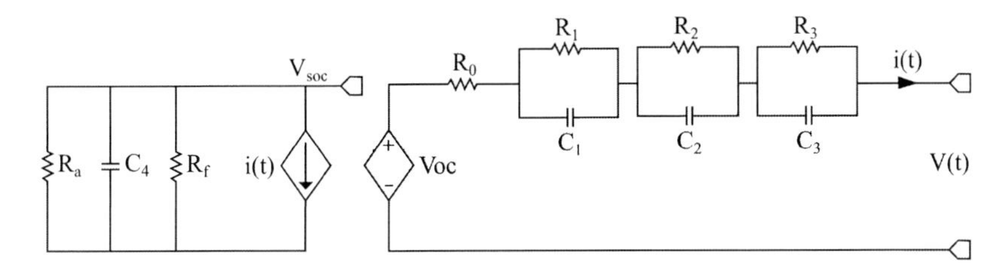

Figure 14. Kroeze and Krein model equivalent circuits.

In order to more precisely determine the battery usable capacity, an analysis of the temperature and number of cycles was performed considering different discharge currents. Therefore, the SOC determination was carried out taking into account the effects described by

$$SOC = SOC_{inicial} + \int_0^t f_1[n_{cycle}] f_2[T(t)] f_3[i(t)] i(t) dt, \tag{27}$$

where: $f_3[i(t)]$ is the capacity correction factor.

The circuit on the left provides the output voltage of the model and each component is a nonlinear function of SOC, given by

$$V(t) = Voc - R_0 i(t) + [R_{transient}(\tau_1, \tau_2, \tau_3)] i(t), \tag{28}$$

where: τ_1, τ_2 e τ_3 are time constants defined by the R_1C_1, R_2C_2 and R_3C_3 products, respectively.

The transient behavior is described as

$$R_{transient}(\tau_1, \tau_2, \tau_3) = R_1 e^{-\frac{1}{\tau_1}t} + R_2 e^{-\frac{1}{\tau_2}t} + R_3 e^{-\frac{1}{\tau_3}t}. \tag{29}$$

The function laws that describes the model components, except for R_1 and C_1, were obtained from [8]. The parameters values were estimated according to the battery type used in the work, and these parameters are related to the R_0, R_1, C_1, R_2 and C_2 components, and were estimated for different discharge currents. The determination of the longer time constant τ_3 is a very time consuming effort, so the values found for the discharge process were also used to model the charge process.

2.5.3. Erdinc Model

Another model based on the Chen and Rincón-Mora is the Erdinc model [15], where a resistance R_c was added to the circuit on the right, to describe the rise in battery resistance that occurs with the raise of cycles number, as can be seen in Figure 15.

This model also takes into account the temperature and capacity fading effects, and according to [16], the output voltage can be expressed as

$$V(t) = Voc - i(t)Zeq + \Delta E(t), \tag{30}$$

where: Zeq is the battery internal impedance and $\Delta E(t)$ is the potential correction term used to compensate the equilibrium potential variation caused by

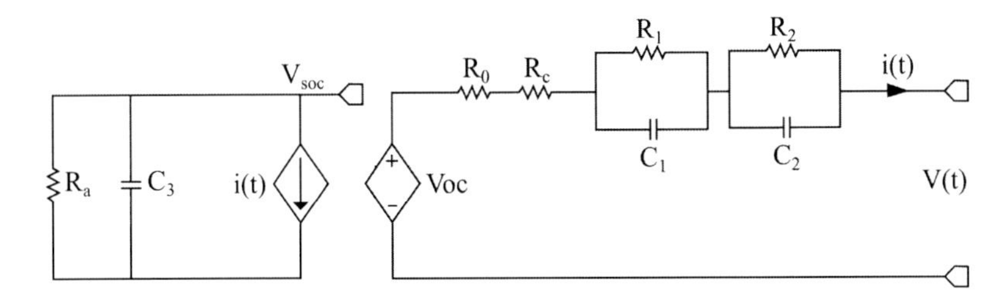

Figure 15. Erdinc model equivalent circuits.

temperature changes [16]. The electrical components related to the output voltage are functions of *SOC*, which is obtained by

$$SOC = SOC_{inicial} - \int \frac{i(t)}{C_{usable}} dt, \tag{31}$$

where: C_{usable} is the battery usable capacity that varies according to the capacity fading.

The capacity fading is the irreversible loss in battery usable capacity that happens over time, because of temperature and the number of cycles. This loss of capacity is associated with the battery degradation and can occur even while the battery is an inactive state (calendar life losses) or when it is active (cycle life losses). These two kinds of losses seem to be linear with time and can drastically increase with the increase in temperature, so it is important to take the temperature into account when modeling the battery capacity fading [15].

The parameter values for the functions were obtained from [8], except for the resistance R_c that is found according to [41] and given by

$$R_c = k_3 N^{1/2}, \tag{32}$$

where: k_3 is the cycle change coefficient and N is the number of cycles [41].

The developed model can be used for battery performance evaluation under different operational conditions, and can be used for the simulations of different systems that include electrical batteries.

2.6. Generic Models

The Generic models are considered accurate battery models, with broad applicability and able to simulate the dynamic behavior of batteries of different technologies under different charge and discharge conditions [35]. An implementation of this model class is the so called Battery model [50], integrated in the SimPowerSystems toolbox for the Matlab/Simulink software package [34].

The internal resistance is considered to be constant and the effective capacity is not influenced by current amplitude changes; the effects of temperature, self-discharge and memory are also neglected [49]. However, for each type of battery there is a specific equation to describe the voltage output for the charge or discharge processes. One example are Lithium Ion Polymer (LiPo) batteries, which are vastly used in mobile devices and their mathematical modeling using the Battery model was studied in [4].

For LiPo batteries, the output voltage $V(t)$, during charge or discharge, is given by the equations (33) and (34), respectively,

$$V(t) = E_0 - Ri - K\left(\frac{Q}{it - 0,1Q}\right)i^* - K\left(\frac{Q}{Q - it}\right)it + Aexp(-Bit), \tag{33}$$

$$V(t) = E_0 - Ri - K\left(\frac{Q}{Q - it}\right)i^* - K\left(\frac{Q}{Q - it}\right)it + Aexp(-Bit), \tag{34}$$

where: E_0 is the constant voltage, K is the polarization constant, i^* is the low frequency dynamic current, it is the battery extracted capacity, Q is the battery maximum capacity, A is the amplitude at the exponential zone, and B is the inverse of the exponential zone time constant [49]. The constant voltage E_0 is obtained by

$$E_0 = V_{Full} + K + Ri - A, \tag{35}$$

where: V_{Full} is the full charge voltage [50]. The polarization constant K is given by

$$K = \frac{(V_{Full} - V_{Nom} + A(exp(-BQ_{Nom}) - 1))(Q - Q_{Nom})}{Q_{Nom}}, \tag{36}$$

where: V_{Nom} is the voltage at the nominal zone end and Q_{Nom} is the capacity at the nominal zone end [50]. Parameters A and B can be obtained from the equations

$$A = V_{Full} - V_{Exp}, \tag{37}$$

$$B = \frac{3}{Q_{Exp}}, \tag{38}$$

where: V_{Exp} is the voltage at the end of the exponential zone and Q_{Exp} is the capacity at the end of the exponential zone [50].

Most of the Battery model parameters are usually present in the battery data sheets, with the exception of three of them (Q_{Nom}, V_{Exp} and Q_{Exp}) that are estimated using discharge curves [34]. The methodology usually adopted for the estimation of these parameters is the visual analysis of specific points in discharge curves obtained by experimental tests [38]. However, some works have explored the use of metaheuristics for the parameters estimation, avoiding the subjective process of visual analysis of the discharge curves [4].

Conclusion

In this chapter, a review of electrical models used in the prediction of batteries charge and discharge behavior was presented. Six electrical model subclasses were presented: Simple models, Thévenin-based models, Impedance-based models, Runtime-based models, Combined models and Generic models, and then main models of each subclass were discussed.

The Simple models cannot capture the most important aspects of the battery dynamics and can hardly be applied in practice. With Thévenin-based models, it is possible to get the battery transient response for a specific SOC state, with a constant Voc, but they are not useful to model the battery lifetime, specially considering variable discharge rates. In contrast, the Impedance-based models can simulate the battery response for variable discharge or charge currents, but since they do not take into account SOC and temperature variations, they are not able to simulate the battery lifetime, and require the EIS method, necessary for the models parametrization, is considered to be complex and unpractical. The main advantage of Runtime-based models is the ability to simulate the battery lifetime, but they only offer good results when constant discharge rates are considered. Combined models can be used to get the battery steady state and transient response with accuracy, providing the battery lifetime simulation. Combined models can also take into account the effects of temperature, number of cycles and battery capacity fading. Generic models can simulate the charge and discharge behavior for different battery technologies, having a wide applicability.

The improvement and combination of different electrical models resulted in the development of new models, capable of better describing the battery behavior, and able to predict, with good accuracy, the battery lifetime and the charge and discharge process dynamics. Generic and Combined models are the most sophisticated electrical models, being accurate while still maintaining a reasonably low complexity.

References

[1] S. M. Alavi, A. Mahdi, S. Payne, and D. A. Howey. Identifiability of generalised randles circuit models. *IEEE Transactions on Control System Technology*, 2016.

[2] J. Appelbaum and R. Weiss. An electrical model of the lead-acid battery. In *Telecommunications Energy Conference, 1982. INTELEC 1982. International*, pages 304–307, Oct 1982.

[3] J. Brand, Z. Zhang, and R. K. Agarwal. Extraction of battery parameters of the equivalent circuit model using a multi-objective genetic algorithm. *Journal of Power Sources*, 247:729 – 737, 2014.

[4] M. F. Brondani, A. Sausen, P. S. Sausen, and M. O. Binelo. Battery model parameters estimation using simulated annealing. *TEMA - Tendências em Matemática Aplicada e Computacional*, 18:127–135, 2017.

[5] S. Buller, E. Karden, D. Kok, and R. W. D. Doncker. Modeling the dynamic behavior of supercapacitors using impedance spectroscopy. *IEEE Transactions on Industry Applications*, 38(6):1622–1626, Nov 2002.

[6] M. A. Casacca and Z. M. Salameh. Determination of lead-acid battery capacity via mathematical modeling techniques. *IEEE Transactions on Energy Conversion*, 7(3):442–446, Sep 1992.

[7] H. L. Chan. A new battery model for use with battery energy storage systems and electric vehicles power systems. In *Power Engineering Society Winter Meeting, 2000. IEEE*, volume 1, pages 470–475, 2000.

[8] M. Chen and G. Rincón-Mora. Accurate electrical battery model capable of predicting runtime and i-v performance. *IEEE Transactions on Energy Conversion*, 21(2):504–511, June 2006.

[9] C. Chiasserini and R. Rao. Pulsed battery discharge in communication devices. *Proceedings of the 5th International Conference on Mobile Computing and Networking*, pages 88–95, 1999.

[10] J. P. Cun, J. N. Fiorina, M. Fraisse, and H. Mabboux. The experience of a ups company in advanced battery monitoring. In *Telecommunications Energy Conference, 1996. INTELEC '96., 18th International*, pages 646–653, Oct 1996.

[11] J. Dambrowski. Validation of impedance-data and of impedance-based modeling approach of electrochemical cells by means of mathematical system theory. In *39th Annual Conference of the IEEE Industrial Electronics Society, 10th-13th November*, 2013.

[12] G. M. de León, C. Sifuentes-Gallardo, A. Moreno-BÃąez, E. García-Domínguez, and R. Magallanes-Quintanar. Time-frequency analysis of a pulsed excitation and its application in randles model. In *Mechatronics, Electronics and Automotive Engineering (ICMEAE), 2015 International Conference on*, pages 157–161, Nov 2015.

[13] M. Doyle, T. F. Fuller, and J. Newman. Simulation and optimization of the dual lithium ion insertion cell. *Journal of the Electrochemical Society*, 141(1):1–10, 1994.

[14] M. Doyle, T. F. Fuller, and J. S. Newman. Modeling of galvanostatic charge and discharge of the lithium/polymer/insertion cell. *Journal of The Electrochemical Society*, 140:1526–1533, 1993.

[15] O. Erdinc, B. Vural, and M. Uzunoglu. A dynamic lithium-ion battery model considering the effects of temperature and capacity fading. In *2009 International Conference on Clean Electrical Power*, pages 383–386, June 2009.

[16] L. Gao, S. Liu, and R. A. Dougal. Dynamic lithium-ion battery model for system simulation. *IEEE Transactions on Components and Packaging Technologies*, 25(3):495–505, Sep 2002.

[17] H. He, R. Xiong, H. Guo, and S. Li. Comparison study on the battery models used for the energy management of batteries in electric vehicles. *Energy Conversion and Management*, 64:113 – 121, 2012. {IREC} 2011, The International Renewable Energy Congress.

[18] H. He, R. Xiong, X. Zhang, F. Sun, and J. Fan. State-of-charge estimation of the lithium-ion battery using an adaptive extended kalman filter

based on an improved thevenin model. *IEEE Transactions on Vehicular Technology*, 60(4):1461–1469, May 2011.

[19] T. Hu and H. Jung. Simple algorithms for determining parameters of circuit models for charging/discharging batteries. *Journal of Power Sources*, 233:14 – 22, 2013.

[20] X. Hu, S. Li, and H. Peng. A comparative study of equivalent circuit models for li-ion batteries. *Journal of Power Sources*, 198:359 – 367, 2012.

[21] F. Huet. A review of impedance measurements for determination of the state-of-charge or state-of-health of secondary batteries. *Journal of Power Sources*, 70(1):59 – 69, 1998.

[22] E. Karden, S. Buller, and R. W. D. Doncker. A method for measurement and interpretation of impedance spectra for industrial batteries. *Journal of Power Sources*, 85(1):72 – 78, 2000.

[23] T. Kim and W. Qiao. A hybrid battery model capable of capturing dynamic circuit characteristics an nonlinear capacity effects. *IEEE Wireless Communications and Networking Conference*, 26:1172–1180, 2011.

[24] Y.-H. Kim and H.-D. Ha. Design of interface circuits with electrical battery models. *IEEE Transactions on Industrial Electronics*, 44(1):81–86, Feb 1997.

[25] R. C. Kroeze and P. T. Krein. Electrical battery model for use in dynamic electric vehicle simulations. In *2008 IEEE Power Electronics Specialists Conference*, pages 1336–1342, June 2008.

[26] E. Kuhn, C. Forgez, P. Lagonotte, and G. Friedrich. Modelling ni-mh battery using cauer and foster structures. *Journal of Power Sources*, 158(2):1490 – 1497, 2006. Special issue including selected papers from the 6th International Conference on Lead-Acid Batteries (LABAT 2005, Varna, Bulgaria) and the 11th Asian Battery Conference (11 ABC, Ho Chi Minh City, Vietnam) together with regular papers.

[27] K. Lahiri, A. Raghunathan, S. Dey, and D. Panigrahi. Battery-driven system design: A new frontier in low power design. *Proc. Intl. Conf. on VLSI Design/ASP-DAC*, pages 261–267, January 2002.

[28] S. Laribi, K. Mammar, M. Hamouda, and Y. Sahli. Impedance model for diagnosis of water management in fuel cells using artificial neural networks methodology. *International Journal of Hydrogen Energy*, 41(38):17093 – 17101, 2016.

[29] F. Leng, C. M. Tan, R. Yazami, and M. D. Le. A practical framework of electrical based online state-of-charge estimation of lithium ion batteries. *Journal of Power Sources*, 255(Complete):423–430, 2014.

[30] S. E. Li, B. Wang, H. Peng, and X. Hu. An electrochemistry-based impedance model for lithium-ion batteries. *Journal of Power Sources*, 258:9 – 18, 2014.

[31] C. Lin, H. Mu, R. Xiong, and W. Shen. A novel multi-model probability battery state of charge estimation approach for electric vehicles using h-infinity algorithm. *Applied Energy*, 166:76 – 83, 2016.

[32] T. Luo, L. Li, V. Ghorband, Y. Zhan, H. Song, and J. B. Christen. A portable impedance-based electrochemical measurement device. In *2016 IEEE International Symposium on Circuits and Systems (ISCAS)*, pages 2891–2894, May 2016.

[33] J. Marcos, A. Lago, C. M. Penalver, J. Doval, A. Nogueira, C. Castro, and J. Chamadoira. An approach to real behaviour modeling for traction lead-acid batteries. In *2001 IEEE 32nd Annual Power Electronics Specialists Conference (IEEE Cat. No.01CH37230)*, volume 2, pages 620–624 vol.2, 2001.

[34] MathWorks. Implement generic battery model, 2016.

[35] S. Mousavi and M. Nikdel. Various battery models for various simulation studies and applications. *Renewable and Sustainable Energy Reviews*, 32:477 – 485, 2014.

[36] S. Pang, J. Farrell, J. Du, and M. Barth. Battery state-of-charge estimation. In *American Control Conference, 2001. Proceedings of the 2001*, volume 2, pages 1644–1649 vol.2, 2001.

[37] D. Panigrahi T, D. Panigrahi, C. Chiasserini, S. Dey, R. Rao, A. Raghunathan, and K. Lahiri. Battery life estimation of mobile embedded sys-

tems. In *VLSI Design, 2001. Fourteenth International Conference on*, pages 57–63, 2001.

[38] C. M. D. Porciuncula, A. T. Z. R. Sausen, and P. S. Sausen. *Mathematical Modeling for Predicting Battery Lifetime through Electrical Models*. Advances in Mathematics Research. Nova Science Publishers Incorporated, 2015.

[39] E. Prada, J. Bernard, and V. Sauvant-Moynot. Ni-mh battery ageing: from comprehensive study to electrochemical modelling for state-of charge and state-of-health estimation. *IFAC Proceedings Volumes*, 42(26):123 – 131, 2009.

[40] D. Rakhmatov and S. Vrudhula. An analytical high-level battery model for use in energy management of portable electronic systems. *National Science Foundation's State/Industry/University Cooperative Research Centers (NSFS/IUCRC) Center for Low Power Electronics (CLPE)*, pages 1–6, 2001.

[41] P. Ramadass, B. Haran, R. White, and B. N. Popov. Mathematical modeling of the capacity fade of li-ion cells. *Journal of Power Sources*, 123(2):230 – 240, 2003.

[42] J. E. B. Randles. Kinetics of rapid electrode reactions. *Discuss. Faraday Soc.*, 1:11–19, 1947.

[43] S. M. Rezvanizaniani, Z. Liu, Y. Chen, and J. Lee. Review and recent advances in battery health monitoring and prognostics technologies for electric vehicle (ev) safety and mobility. *Journal of Power Sources*, 256:110 – 124, 2014.

[44] L. C. Romio, A. T. Z. R. Sausen, P. S. Sausen, and M. Reimbold. *Mathematical Modeling of the Lithium-ion battery Lifetime using System Identification Theory*. Advances in Mathematics Research. Nova Science Publishers Incorporated, 2015.

[45] Z. M. Salameh, M. A. Casacca, and W. A. Lynch. A mathematical model for lead-acid batteries. *IEEE Transactions on Energy Conversion*, 7(1):93–98, Mar 1992.

[46] A. Seaman, T.-S. Dao, and J. McPhee. A survey of mathematics-based equivalent-circuit and electrochemical battery models for hybrid and electric vehicle simulation. *Journal of Power Sources*, 256:410 – 423, 2014.

[47] C. M. Shepherd. Design of primary and secondary cells. *Journal of The Electrochemical Society*, 112:657–664, 1965.

[48] B. Suthar, V. Ramadesigan, P. W. C. Northrop, R. B. Gopaluni, S. Santhanagopalan, R. D. Braatz, and V. R. Subramanian. Optimal control and state estimation of lithium-ion batteries using reformulated models. In *American Control Conference, ACC 2013, Washington, DC, USA, June 17-19, 2013*, pages 5350–5355, 2013.

[49] O. Tremblay and L.-A. Dessaint. Experimental validation of a battery dynamic model for ev applications. *World Electric Vehicle Journal*, 3:289–298, 2009.

[50] O. Tremblay, L.-A. Dessaint, and A.-I. Dekkiche. A generic battery model for the dynamic simulation of hybrid electric vehicles. In *Vehicle Power and Propulsion Conference, 2007. VPPC 2007. IEEE*, pages 284–289, Sept 2007.

[51] U. TrÃűltzsch, O. Kanoun, and H.-R. TrÃd'nkler. Characterizing aging effects of lithium ion batteries by impedance spectroscopy. *Electrochimica Acta*, 51(8âĂŞ9):1664 – 1672, 2006. Electrochemical Impedance SpectroscopySelection of papers from the 6th International Symposium (EIS 2004) 16-21 May 2004, Cocoa Beach, FL, {USA}.

[52] J. Xu, C. C. Mi, B. Cao, and J. Cao. A new method to estimate the state of charge of lithium-ion batteries based on the battery impedance model. *Journal of Power Sources*, 233:277 – 284, 2013.

[53] N. Yang, X. Zhang, and G. Li. State-of-charge estimation for lithium ion batteries via the simulation of lithium distribution in the electrode particles. *Journal of Power Sources*, 272:68–78, 2014.

[54] C. Zhang, K. Li, S. Mcloone, and Z. Yang. Battery modelling methods for electric vehicles - a review. In *2014 European Control Conference (ECC)*, pages 2673–2678, June 2014.

[55] H. Zhang and M.-Y. Chow. Comprehensive dynamic battery modeling for phev applications. In *IEEE PES General Meeting*, pages 1–6, July 2010.

In: Advances in Mathematics Research
Editor: Albert R. Baswell
ISBN: 978-1-53612-512-2

Chapter 5

AN ACCURATE MODELING AND PERFORMANCE OF MULTISTAGE LAUNCH VEHICLES FOR MICROSATELLITES VIA A FIREWORK ALGORITHM

M. Pontani*, M. Pallone† and P. Teofilatto‡
Department of Astronautical, Electrical and Energy Engineering,
University of Rome "La Sapienza", Rome, Italy

Abstract

Multistage launch vehicles of reduced size, such as "Super Strypi" or "Sword", are currently investigated for the purpose of providing launch opportunities for microsatellites. This work proposes a general methodology for the accurate modeling and performance evaluation of launch vehicles dedicated to microsatellites. For illustrative purposes, the approach at hand is applied to the Scout rocket, a micro-launcher used in the past. Aerodynamics and propulsion are modeled with high fidelity through interpolation of available data. Unlike the original Scout, the terminal optimal ascent path is determined for the upper stage, using a firework algorithm in conjunction with the Euler-Lagrange equations and the Pontryagin minimum principle. Firework algorithms represent a

*Corresponding Author Email: mauro.pontani@uniroma1.it.
†Corresponding Author Email: pallone.1420138@studenti.uniroma1.it.
‡Corresponding Author Email: paolo.teofilatto@uniroma1.it.

recently-introduced heuristic technique, not requiring any starting guess and inspired by the firework explosions in the night sky. The numerically results prove that this methodology is easy-to-implement, robust, precise and computationally effective, although it uses an accurate aerodynamic and propulsive model.

Keywords: multistage launch vehicles, trajectory optimization, firework algorithm optimization

1. Introduction

Currently, microsatellites can be launched according to the time and orbital requirements of a main payload. The limited costs of microsatellites and their capability to be produced and ready for use in short time make them particularly suitable to face an emergency (responsive space), therefore small launch vehicles dedicated to microsatellites would be very useful. In order to reduce the launcher size without increasing too much the launch cost per kg of payload it is necessary to simplify the launch system as much as possible, including the guidance algorithms.

In general, the numerical solution of aerospace trajectory optimization problems is not trivial and has been pursued with different approaches in the past. However, only a relatively small number of publications are concerned with trajectory optimization of multistage launch vehicles [10, 11, 3, 5, 4, 7, 6, 19, 9, 16, 1, 15]. Calise [3] and Gath [4] proposed and applied a hybrid analytic/numerical approach, based on homotopy and starting with the generation of the optimal solution in a vacuum. They adopted the approximate linear gravity model, and the same did Lu [7] and [6], who applied a multiple-shooting method to optimizing exoatmospheric trajectories composed of two powered phases separated by a coast arc. Weigel [19] used a similar indirect, multiple-shooting approach to analyze and optimize the ascent trajectories of two launch vehicles with splashdown constraints. Miele [11] developed and applied the indirect multiple-subarc gradient restoration algorithm to optimizing the two-dimensional ascending trajectory of a three-stage rocket in the presence of dynamical and control constraints. The previously cited works [11, 3, 4, 7, 6, 19] make use of indirect algorithms and require a considerable deal of effort for deriving the analytical conditions for optimality and for the subsequent programming and debugging. Furthermore, these methods can suffer from a slow rate of convergence and an excessive dependence on the starting guess. This difficulty has been occasionally circumvented through homotopy [3, 4, 9], but this adds further complexity to the solution process. Other papers deal with direct numerical techniques applied to multistage rocket trajectory optimization. Roh [16] used a collocation method for optimizing the performance of a four-stage rocket, whose two-dimensional trajectory was assumed to be composed of three thrust phases and a coast arc of specified duration. Collocation was also employed by Jamilnia [5], with the additional task of determining the optimal staging, and by Martinon [9], for the purpose of validating the numerical results attained through indirect shooting. This latter paper refers to the Ariane V launch vehicle and is specifically devoted to investigating singular arcs.

The work that follows is concerned with a novel approach, which is intended to supply a fast performance evaluation for multistage rockets with given characteristics, under some simplifying assumptions regarding the final orbit. The aerodynamic and propulsive model used in this work is very accurate and it was obtained through previous numerical computation and using the real thrust curve for each stage. Although the model is very precise, this adds a very little amount of computational effort but increase the accuracy of the results. Programming, debugging, and testing the algorithmic codes is very easy, as existing routines are used, in conjunction with analytical developments and a simple implementation of firework algorithm. The technique described in this work is applied to a four-stage rocket, whose two-dimensional trajectory is composed of the following thrust phases and coast arcs:

1. first stage propulsion
2. second stage propulsion
3. third stage propulsion
4. coast arc (after the third stage separation)
5. fourth stage thrust phase

where the first three stages are guided maintaining constant the aerodynamic angle of attack α. This law is intended to remain under the maximum allowed aerodynamic load ($Q\alpha$), and this grants the simplicity of the guidance for microsatellite dedicated launchers. In general, the inclusion of a coast arc (between two powered phases) leads to substantial propellant savings and this circumstance justifies the trajectory structure assumed in this research. Usually the coast duration increases as the injection altitude increases, as remarked by Lu [6]. The total efficiency of the launcher often depends on the efficiency of the last stage so for the fourth stage an optimal guidance is used. In the last stage thrust phase the problem of minimizing the propellant is solved defining a Hamiltonian function which is minimized through the Pontryagin minimum principle. The optimization algorithm used in this work is the firework algorithm [17], a novel swarm intelligence algorithm inspired by the explosions of fireworks in a night sky. The concept that underlies this method is relatively simple: a firework explodes in the search space of the unknown parameters, with amplitude and number of sparks determined dynamically. The initial fireworks are generated randomly, and the succeeding iterations preserve the best sparks. The novelty of this manuscript with respect to the previous works [13, 12] lies on the very accurate propulsive and aerodynamic models that in the other works are maintained constants for the sake of simplicity and computational time. Furthermore the use of a novel heuristic algorithm is highlighted. Definitely, the methodology treated in this paper is intended to yield a near-optimal ascent path useful for performance evaluation of multistage rockets, with accurate aerodynamic and propulsive modeling.

2. Rocket Modeling

The four stage rocket that is being considered is similar to the Scout launcher, which is a rocket designed to place small satellites into low orbit. It has specified structural, propulsive and aerodynamic characteristics and it is represented in Fig. 1.

The mass distribution of the launch vehicle can be described in terms of masses of subrockets: subrocket 1 is the entire rocket with all the four stages, subrocket 2 is the launch vehicle after the separation of the first stage, subrocket 3 is the launch vehicle after the separation of the first two stages and subrocket 4 is represented by the last stage only. Let $m_0^{(i)}$ denote the initial mass of the subrocket i, this mass is composed of a structural mass $m_S^{(i)}$, a propellant mass $m_P^{(i)}$ and a payload mass $m_U^{(i)}$:

$$m_0^{(i)} = m_S^{(i)} + m_P^{(i)} + m_U^{(i)} \tag{1}$$

For the first three subrockets $m_U^{(i)}$ (i=1,2,3) coincides with the initial mass of the

Figure 1. Scout rocket geometry.

subsequent subrocket (i.e. $m_U^{(i)} = m_0^{(i+1)}$). With regard to the fourth stage, its initial mass is specified by using the data reported in the Scout manual. For final altitudes of 300 and 1700 km, this manual reports a payload mass of 200 and 50 kg, respectively. These two masses correspond to the overall mass distributions summarized in Tables 2 and 1, which therefore refer to two distinct initial conditions at launch. It is apparent that any propellant saving for the upper stage translates into a possible increase of the payload mass. In fact, minimizing the propulsion duration implies minimizing the propellant needed to operate the upper stage, and this portion of unnecessary propellant mass can be replaced by additional payload mass.

Table 1. Mass distribution for the first three subrockets (circular target orbit at 1700 km of altitude)

i	$m_0^{(i)}$	$m_S^{(i)}$	$m_P^{(i)}$	$m_U^{(i)}$
1	21493 kg	1736 kg	12810 kg	6747 kg
2	6747 kg	915 kg	3749 kg	1883 kg
3	1883 kg	346 kg	1173 kg	364 kg

Table 2. Mass distribution for the first three subrockets (circular target orbit at 300 km of altitude)

i	$m_0^{(i)}$	$m_S^{(i)}$	$m_P^{(i)}$	$m_U^{(i)}$
1	21643 kg	1736 kg	12810 kg	6897 kg
2	6897 kg	915 kg	3749 kg	2033 kg
3	2033 kg	346 kg	1173 kg	514 kg

2.1. Propulsive Thrust

The propulsive characteristic of the launch vehicle can be described in terms of thrust magnitude $T^{(j)}$ and specific impulse $I_{SP}^{(j)}$, with superscript j referring to the stage number. The specific impulses for the four stages are listed in Table 3.

Table 3. Specific impulse for the four stages

j	1	2	3	4
$Isp^{(j)}$	260 s	288 s	284 s	270 s

While the specific impulse is considered time-independent for all the four stages, the thrust is obtained through a linear interpolation of the experimental thrust data which are given at discrete times. Fig. 2 portrays the thrust curves for each motor, whose burnout time is $t_{B(j)}$.

2.2. Aerodynamics

Aerodynamic modeling is composed of two steps:

1. derivation of C_D and C_L at a relevant number of Mach numbers and angles of attack
2. fourth degree polynomial interpolation of the aerodynamic parameters with Mach number and angle of attack

Following the approach presented by Mangiacasale [8], the aerodynamics of the Scout rocket was modeled through the Missile DATCOM software [2] for the first three subrockets. The aerodynamic force is assumed to be composed of

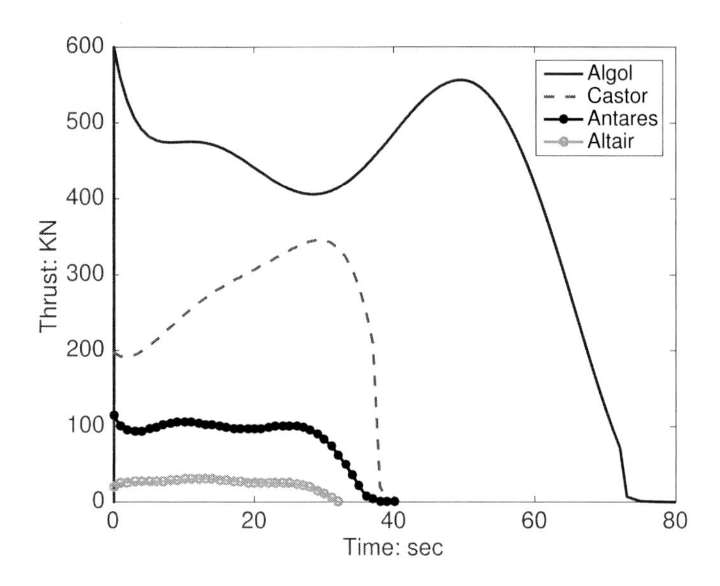

Figure 2. Thrust curve for the four stages.

two components: lift force L and drag force D. Given the aerodynamic surface S, the atmospheric density ρ, the speed relative to the Earth atmosphere v, and the lift and drag coefficients, C_L and C_D respectively, the two components are:

$$L = \frac{1}{2} C_L(\alpha, M) S^{(i)} \rho v^2 \text{ and } D = \frac{1}{2} C_D(\alpha, M) S^{(i)} \rho v^2 \tag{2}$$

where the coefficients C_L and C_D depend on the Mach number M and the aerodynamic angle of attack α. The aerodynamic surfaces used in the computation are the cross surfaces and are listed in Table 4. The C_L and C_D coefficients are

Table 4. Aerodynamic surfaces for the first three subrockets

j	1	2	3
$S^{(i)}$	1.026 m^2	0.487 m^2	0.458 m^2

obtained using a four degree polynomial interpolation to speed up the computation time. The polynomial expression (valid both for C_L and C_D) is:

$$\begin{aligned} C_k = {} & C_{00,k} + C_{10,k}\alpha + C_{01,k}M + C_{20,k}\alpha^2 \\ & + C_{11,k}\alpha M + C_{02,k}M^2 + C_{30,k}\alpha^3 \\ & + C_{21,k}\alpha^2 M + C_{12,k}\alpha M^2 + C_{03,k}M^3 \\ & + C_{40,k}\alpha^4 + C_{31,k}\alpha^3 M + C_{22,k}\alpha^2 M^2 \\ & + C_{13,k}\alpha M^3 + C_{04,k}M^4 \end{aligned} \tag{3}$$

with $k = L, D$. The polynomial coefficients are determined through a least square approach that compares the Datcom values with the interpolated values. To improve the quality of the interpolation, the distinction between subsonic ($M \in [0, 0.8]$), transonic ($M \in [0.8, 1.2]$) and supersonic ($M \in [1.2, 10]$) polynomial coefficients has been done. While the subsonic and the supersonic cases follow the Eq. (3), in the transonic case an embedded Matlab interpolation routine was used due to the abrupt behavior of the coefficients as M varies around 1. Table 5 and Table 6 report the coefficients used for each subrocket and an example of the interpolation is in Fig. 3. The fourth stage usually flies over 120 km of altitude so the atmosphere is very rarefied and is neglected.

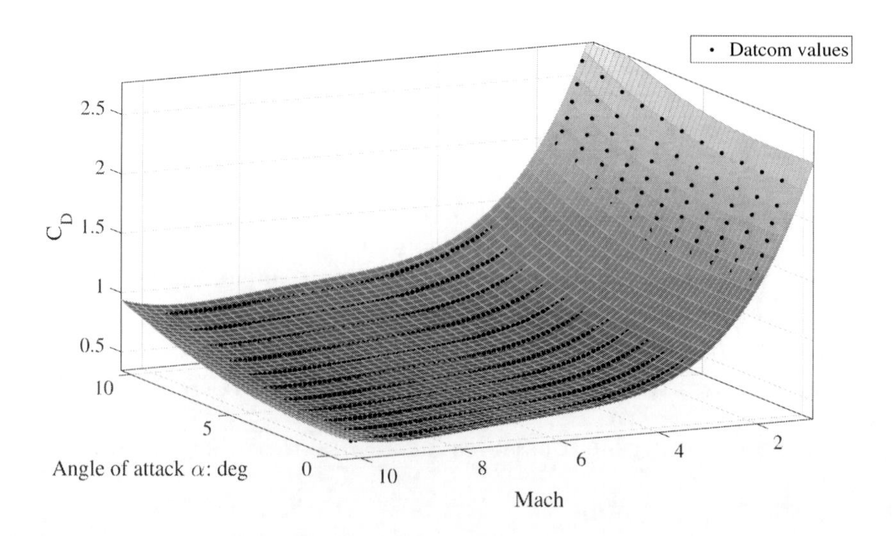

Figure 3. Fourth degree polynomial interpolation surface for the C_D of the first subrocket in supersonic flight.

Table 5. Polynomial coefficients for aerodynamic interpolation, subsonic case

	subsonic					
	subrocket 1		subrocket 2		subrocket 3	
	C_D	C_L	C_D	C_L	C_D	C_L
C_{00}	0.213	0.141	0.160	0.004	0.113	0.002
C_{10}	-0.306	5.638	-0.008	1.714	0.015	1.826
C_{01}	0.017	-1.857	0.961	-0.041	1.055	-0.020
C_{20}	7.398	6.636	1.959	9.338	1.908	3.333
C_{11}	2.781	10.440	0.092	0.294	-0.114	-0.237
C_{02}	-0.073	6.886	-3.484	0.120	-3.828	0.061
C_{30}	5.448	-6.733	7.732	-2.092	2.473	2.403
C_{21}	-4.116	-6.103	-0.330	-3.803	0.370	-2.110
C_{12}	-6.230	-26.20	-0.074	0.373	0.327	1.569
C_{03}	-0.361	-9.520	3.804	-0.132	4.25	-0.070
C_{40}	-9.063	-32.270	-2.443	-7.244	-1.494	-15.90
C_{31}	-1.953	14.090	3.628	6.947	2.080	2.065
C_{22}	5.965	2.232	-0.517	5.463	-1.142	3.364
C_{13}	3.866	19.890	-0.008	-1.335	-0.229	-2.122
C_{04}	0.769	4.373	-0.492	0.048	-0.638	0.027

3. Rocket Dynamics

As the rocket performance is being evaluated, the simulations are performed in the most favorable dynamical conditions, i.e. equatorial trajectory and launch toward the East direction. The four-stage rocket is modeled as a point mass, in the context of a two-degree-of-freedom problem.

The rocket motion is described more easily in a rotating (i.e. non inertial) reference frame. The *Earth Centered Earth Fixed* (ECEF) reference frame represents a reference system that rotates with the Earth and has the origin in its center. The ECEF system rotates with a speed $\omega_E = 7.292115 \times 10^{-5}\ s^{-1}$ with respect to an inertial Earth-centered frame (ECI), denoted with $(\hat{c}_1, \hat{c}_2, \hat{c}_3)$. Both frames share the same origin O. $\hat{c}_1$ is the vernal axis and the vector $\hat{c}_3 = \hat{k}$ is aligned with the planet rotation axis and is positive northward. Therefore $\omega_E \hat{k}$ represents the (vector) rotation rate of the ECEF frame with respect to the

Table 6. Polynomial coefficients for aerodynamic interpolation, supersonic case

	supersonic					
	subrocket 1		subrocket 2		subrocket 3	
	C_D	C_L	C_D	C_L	C_D	C_L
C_{00}	6.024	0.098	4.284	-0.515	4.814	-0.381
C_{10}	0.810	22.290	1.382	9.774	0.862	4.495
C_{01}	-4.449	-0.202	-2.479	0.608	-2.855	0.466
C_{20}	-5.188	-98.280	-9.404	-86.950	-5.682	-39.830
C_{11}	-0.250	-7.475	-0.942	-4.251	-0.583	-1.072
C_{02}	1.481	0.114	0.815	-0.260	0.974	-0.203
C_{30}	56.290	422.40	55.090	443.40	31.140	201.0
C_{21}	5.577	41.750	3.890	34.230	2.599	15.590
C_{12}	-0.031	1.059	0.201	0.705	0.132	0.044
C_{03}	-0.227	-0.024	-0.120	0.048	-0.155	0.037
C_{40}	-50.640	-656.60	-49.710	-765.50	-36.330	-353.70
C_{31}	-3.315	-51.920	-3.605	-47.130	-1.794	-20.810
C_{22}	-0.608	-3.569	-0.323	-2.867	-0.263	-1.344
C_{13}	0.009	-0.043	-0.014	-0.029	-0.010	0.012
C_{04}	0.013	0.002	0.007	-0.003	0.009	-0.002

ECI frame. The unit vector $\hat{\imath}$ intersects the Greenwich meridian at all times. The ECEF-frame is associated with $(\hat{\imath}, \hat{\jmath}, \hat{k})$, which form a right-handed, time-dependent sequence of unit vectors. As the reference Greenwich meridian rotates with rotation rate ω_E, its angular position (with respect to the ECI-frame) is identified by its absolute longitude (usually termed Greenwich sidereal time) $\theta_g(t) = \theta_g(\bar{t}) + \omega_E(t - \bar{t})$, where $\bar{t}$ denotes a generic time instant. The position vector of the orbiting spacecraft in the ECEF-frame is denoted with $\mathbf{r}$, whereas the subscript *I* corresponds to a quantity in the ECI-frame. The inertial velocity is related to the (relative) velocity $\mathbf{v}$ through the following expression:

$$\mathbf{v}_I = \mathbf{v} + \omega_E \times \mathbf{r} \tag{4}$$

This means that, unlike the position vector, the velocity vector in rotating coordinates, $\mathbf{v}$, does not coincide with the inertial velocity vector, $\mathbf{v}_I$. As the entire trajectory lies in the equatorial plane, the flight path angle γ suffices to identify the velocity direction. The instantaneous position is defined through r (=$|\mathbf{r}|$) and the geographical longitude ξ. From inspection of Fig 4, it is apparent that

$$\mathbf{v} = v[\sin\gamma, \cos\gamma][\hat{r}, \hat{E}]^T \tag{5}$$

$$\mathbf{r} = r[\cos\xi, \sin\xi][\hat{r}, \hat{E}]^T \tag{6}$$

The overall aerodynamic force $\mathbf{A}$ is conveniently written in the $(\hat{n}, \hat{v}, \hat{h})$ frame (with $\hat{v}$ aligned with $\mathbf{v}$) as the sum of the lift and drag forces:

$$\mathbf{A} = L\hat{n} - D\hat{v} \tag{7}$$

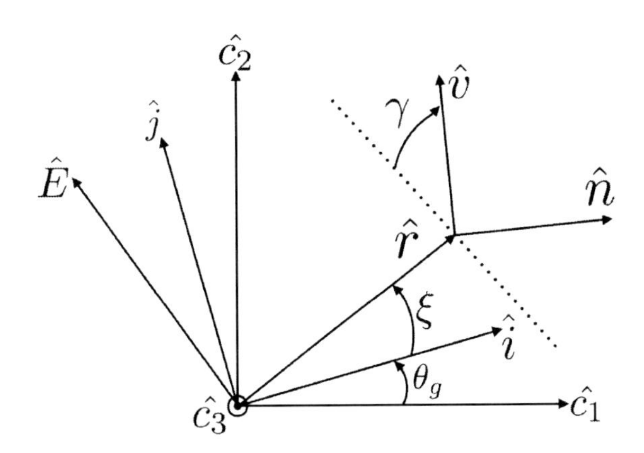

Figure 4. $(\hat{r}, \hat{E})$ frame, $(\hat{n}, \hat{v})$ frame and related angles.

3.1. Equations of Motion

The equations of motion that govern the two dimensional rocket dynamics can be conveniently written in terms of its radius r, flight-path angle γ, velocity v and mass m. These variables refer to the relative motion in an Earth-centered rotating frame (ECEF). They form the state vector $\mathbf{x}_R$ $([r\ \gamma\ v\ m]^T)$ of the launch vehicle (in rotating coordinates). Omitting the superscript i, for each subrocket the equations of motion are:

$$\dot{r} = v\sin\gamma \tag{8}$$

$$\dot{\gamma} = \frac{T}{mv}\sin\alpha_T + \left(\frac{v}{r} - \frac{\mu_E}{r^2 v}\right)\cos\gamma + \frac{L}{mv} + 2\omega_E + \frac{\omega_E^2 r}{v}\cos\gamma \tag{9}$$

$$\dot{v} = \frac{T}{mv}\cos\alpha_T - \frac{\mu_E}{r^2}\sin\gamma - \frac{D}{m} + \omega_E^2 r\sin\gamma \tag{10}$$

$$\dot{m} = -\frac{T}{I_{sp}g_0} \tag{11}$$

where α_T refers to the thrust angle, μ_E (=398600.4 km^3/sec^2) is the Earth gravitational parameter. As the thrust vector is assumed to be coplanar with the position vector $\mathbf{r}$ and the velocity vector $\mathbf{v}$, the angle α_T suffices to define its direction, which is taken clockwise from $\mathbf{v}$. With the exception of *m*, the state $\mathbf{x}_R$ is continuos across stage separations, which occur at time t_{b1} (first stage separation), t_{b2} (second stage separation) and t_{b3} (third stage separation). The fourth stage trajectory is assumed to be composed of two phases: a coast arc and a thrust phase. During the coast arc, the true anomaly variation Δf suffices to describe the rocket dynamics. In fact, if t_{CO} represents the ignition time of the fourth stage, then $f_4 \equiv f(t_{CO}) = f(t_{b3}) + \Delta f$. The orbital elements a and e do not vary during the coast arc. Hence, they can be computed at separation of the third stage through the following steps:

1. derivation of the inertial state variables (r_I, γ_I, v_I) from the relative state variables (r, γ, v)

2. derivation of the orbital elements (a, e, f) from the inertial state variables (r_I, γ_I, v_I)

The velocity vectors $\mathbf{v}_I$ and $\mathbf{v}$ have an expression similar to Eq. 5 in the rotating frame $(\hat{r}, \hat{E}, \hat{N})$. Due to this and the fact that $\omega_E \times \mathbf{r} = \omega_E r\hat{E}$, Eq. (4) yields to two simple relations,

$$v_I = \sqrt{v^2 + (\omega_E r)^2 + 2v\omega_E r\cos\gamma} \tag{12}$$

$$\gamma_I = \arcsin\frac{v\sin\gamma}{v_I} \tag{13}$$

With regard to step (2), the in plane orbital elements (a, e, f) can be easily calculated from r, v_I and γ_I ([14]). In fact, the conservation of energy yields the semimajor axis:

$$a = \frac{\mu_E r}{2\mu_E - r v_I^2} \tag{14}$$

Then using the definition of the magnitude of the angular momentum in terms of orbital elements $h = \sqrt{\mu_E a(1-e^2)}$ and noticing that $h = r v_I \cos\gamma_I$, the eccentricity can be expressed as,

$$e = \sqrt{1 - \frac{r v_I \cos\gamma_I}{\mu_E a}} \tag{15}$$

The true anomaly f can be obtained using the polar equation of the ellipse,

$$r = \frac{a(1-e^2)}{1+e\cos f} \rightarrow \cos f = \frac{a(1-e^2)-r}{re} \tag{16}$$

in conjunction with the radial component of velocity,

$$\sqrt{\frac{\mu_E}{a(1-e^2)}} e \sin f = v \sin\gamma \rightarrow \sin f = \frac{v\sin\gamma}{e}\sqrt{\frac{a(1-e^2)}{\mu_E}} \tag{17}$$

These steps are employed to calculate the true anomaly f_3 at the third stage burnout. So the inertial radius, velocity and the flight path angle at t_{CO} are given by,

$$r(t_{CO}) = \frac{a_3(1-e_3^2)}{1+e_3\cos f_4} \tag{18}$$

$$v_I(t_{CO}) = \sqrt{\frac{\mu_E}{a_3(1-e_3^2)}}\sqrt{1+e_3^2+2e_3\cos f_4} \tag{19}$$

$$\gamma_I(t_{CO}) = \arctan\frac{e_3\sin f_4}{1+e_3\cos f_4} \tag{20}$$

where a_3 and e_3 are the semimajor axis and eccentricity at t_{B3}. During the propulsion phase, the fourth stage motion can be described through the use of the equations that regard r, v_I and γ_I, using the initial conditions reported in (18)-(20). These are identical to Eqs. 8-11 with inertial state variables and aerodynamic and non inertial terms neglected (D=0, L=0, ω_E=0 are introduced in Eqs.(9)-(10)).

4. Performance Evaluation

The control laws are determined using two distinct approaches for the first three stages and the last stage.

4.1. Formulation of the Problem

The desired orbit is assumed to be a circular orbit of radius R_d, therefore

$$r_d(t_f) = R_d \quad v_{I,d}(t_f) = \sqrt{\frac{\mu_E}{R_d}} \quad \gamma_{I,d}(t_f) = 0 \tag{21}$$

The objective is minimizing fuel consumption for the upper stage while injecting the spacecraft into the desired orbit. Hence the objective function for the entire rocket optimization is:

$$J = t_f - t_{CO} \tag{22}$$

Minimizing J implies minimizing the fuel used in the fourth stage propulsive phase so the payload mass can be increased by the same amount of propellant mass saved. The determination of the trajectory of the first three stages is based on maintaining constant the angle of attack during the flight, except for the first 5 seconds of flight of the first subrocket when the thrust direction is radial. The optimal control strategy for the last stage is instead obtained through minimizing the Hamiltonian function.

4.2. Method of Solution

As remarked before, the angle of attack of the first three stages is constant and the thrust angle is assumed to be aligned with the rocket longitudinal axis, thus:

$$\alpha_{T,i} = \alpha_i \text{ (i=1,2,3)} \tag{23}$$

Without this simplification, the use of analytical conditions would lead to several complexities, such as obtaining the partial derivatives with respect to altitude and velocity. To obtain the fourth stage optimal control law, the following optimization problem is defined: find the optimal $\alpha_T(t)$ and the optimal true anomaly f_4 such that J is minimized. The ignition time t_{CO} is computed through the Kepler's law,

$$\begin{aligned} t_{CO} = t_{b3} + \sqrt{\frac{a_3^3}{\mu_E}} [E(t_{CO}) - E(t_{b3}) - e_3 \sin E(t_{CO}) \\ + e_3 \sin E(t_{B3})] \end{aligned} \tag{24}$$

where $E(t_{b3})$ and $E(t_{CO})$ are the eccentric anomalies associated with $f(t_{b3})$ and $f(t_{CO})$. Letting $\mathbf{x} = [x_1, x_2, x_3]^T = [r, v_I, \gamma_I]^T$, to obtain necessary conditions for an optimal solution, a Hamiltonian H and a function of the boundary condition Φ are introduced as

$$\begin{aligned} H \equiv \lambda_1 x_2 \sin x_3 + \lambda_2 \left[\frac{T}{m} \frac{\cos \alpha_T}{x_2} - \frac{\mu_E}{x_1^2} \sin x_3 \right] \\ \lambda_3 \left[\frac{T}{m} \frac{\sin \alpha_T}{x_2} + \left(\frac{x_2}{x_1} - \frac{\mu_E}{x_1^2 x_2} \right) \cos x_3 \right] \end{aligned} \tag{25}$$

$$\begin{aligned} \Phi \equiv (t_f - t_{CO}) + \nu_1 \left[x_{10} - \frac{a_3(1-e_3^2)}{1 + e_3 \cos f_4} \right] \\ + \nu_2 \left[x_{20} - \sqrt{\frac{\mu_E}{a_2(1-e_3^2)}} \sqrt{1 + e_3^2 + 2 e_3 \cos f_4} \right] \\ + \nu_3 \left[x_{30} - \arctan \frac{e_3 \sin f_4}{1 + e_3 \cos f_4} \right] + \nu_4 \left[x_{1f} - R_d \right] \\ + \nu_5 \left[x_{2f} - \sqrt{\frac{\mu_E}{R_d}} \right] + \nu_6 x_{3f} \end{aligned} \tag{26}$$

where $x_{k0} = x_k(t_{CO})$ and $x_{kf} = x_k(t_f)$ (k=1,2,3); $\boldsymbol{\lambda}$ ($\equiv [\lambda_1, \lambda_2, \lambda_3]^T$) and $\boldsymbol{\nu}$ ($\equiv [\nu_1, \nu_2, \nu_3, \nu_4, \nu_5, \nu_6]^T$) represent, respectively, the adjoint variable conjugate to the dynamics Eq. (8)-(11) and to the boundary conditions (Eq. (21)). The necessary conditions for optimality yield the following adjoint equations for the costate $\boldsymbol{\lambda}$:

$$\dot{\lambda_1} = (x_2 \lambda_3 \cos x_3) \frac{1}{x_1^2} - \left(2\mu_E \lambda_2 \sin x_3 + 2\mu_E \lambda_3 \frac{\cos x_3}{x_2} \right) \frac{1}{x_1^3} \tag{27}$$

$$\dot{\lambda_2} = -\lambda_1 \sin x_3 - \lambda_3 \left[\cos x_3 \left(\frac{1}{x_1} + \frac{\mu_E}{x_1^2 x_2^2} \right) - \frac{T}{m x_2^2} \sin \alpha_T \right] \tag{28}$$

$$\dot{\lambda_3} = -x_2 \lambda_1 \cos x_3 + \mu_E \lambda_2 \frac{\cos x_3}{x_1^2} + \lambda_3 \sin x_3 \left(\frac{x_2}{x_1} - \frac{\mu_E}{x_1^2 x_2} \right) \tag{29}$$

in conjunction with the respective boundary conditions,

$$\lambda_{k0} = -\nu_k \;\; and \;\; \lambda_{kf} = \nu_{k+3} \;\; (k = 1, 2, 3) \tag{30}$$

In the presence of initial conditions depending on the parameter f_4 a pair of additional necessary conditions must hold,

$$\frac{\partial \Phi}{\partial f_4} = 0 \text{ and } \frac{\partial^2 \Phi}{\partial f_4^2} \geq 0 \tag{31}$$

The first equation yields a relation that expresses λ_{30} as a function of λ_{10}, λ_{20} and f_4,

$$\begin{aligned}\lambda_{30} = \lambda_{20} & \frac{\sin f_4 \sqrt{1 + e_3^2 + 2e_3 \cos f_4}}{e_3 + \cos f_4} \sqrt{\frac{\mu_E}{a_3(1 - e_3^2)}} \\ & - \lambda_{10} \frac{a_3(1 - e_3^2)}{e_2 + \cos f_4} \frac{1 + e_3^2 + 2e_3 \cos f_4}{(1 + e_3 \cos f_4)^2} \sin f_4\end{aligned} \tag{32}$$

In addition, the optimal control α_T^* can be expressed as a function of the costates through the Pontryagin minimum principle:

$$\alpha_T^* = \arg \min_{\alpha_T} H \tag{33}$$

implying

$$\sin \alpha_T^* = -\frac{\lambda_3}{x_2} \left[\left(\frac{\lambda_3}{x_2} \right)^2 + \lambda_2^2 \right]^{-1/2} \tag{34}$$

and

$$\cos \alpha_T^* = -\lambda_2 \left[\left(\frac{\lambda_3}{x_2} \right)^2 + \lambda_2^2 \right]^{-1/2} \tag{35}$$

Lastly, as the final time is unspecified, the following transversality condition must hold:

$$H(t_f) + \frac{\partial \Phi}{\partial t_f} = 0 \tag{36}$$

implying

$$\frac{n_0^{(4)}}{1 - n_0^{(4)}(t_f - t_{CO})/(g_0 I_{sp}^{(4)})} \sqrt{\left[\frac{\lambda_{3f}}{x_{2f}} \right]^2 - \lambda_{2f}^2} - 1 = 0 \tag{37}$$

where $n_0^{(4)}$ is the initial T/m for the fourth stage. The necessary conditions for optimality allow translating the optimal control problem into a two-point boundary-value problem involving Eqs. (27)-(37), with unknowns represented by the initial values of $\boldsymbol{\lambda}$, f_4 and t_f. The equality constraints reduce the search space where the solution can be located. Moreover, [13] demonstrated that for the problem at hand the transversality condition (Eq.37) can be neglected and replaced by

$$H(t_f) < 0 \tag{38}$$

Therefore, the optimal control α_T^* can be determined without considering the transversality condition, which is in fact ignorable in this context and this improves the performances of the firework algorithm.

In short, the following parameter set can be employed in the solution process: $\{\alpha_1, \alpha_2, \alpha_3, \lambda_{10}, \lambda_{20}, f_4, t_f\}$. The remaining parameter λ_{30} can be easily obtained by means of Eq. (32). Specifically, the technique is based on the following points:

1. given the initial conditions at launch and the values of the angles of attack (constant) at each stage $\{\alpha_1, \alpha_2, \alpha_3\}$, the state Eqs. (8)-(11) are integrated numerically for each subrocket until the third stage burnout time
2. the coast arc is computed analytically using Eqs. (18)-(20)
3. for the upper stage the control variable is expressed as a function of the costate through the Eqs. (34)-(35)
4. the value of λ_{30} is calculated by means of Eqs. (32), after picking the unknown values of the remaining Lagrange multipliers at the initial time (λ_{10} and λ_{20}), and the true anomaly f_4
5. the equations of motion are used together with Eqs. (27)-(29). The respective initial conditions are known once the parameters f_4 (for the state equations) and $\{\lambda_{10}, \lambda_{20}, \lambda_{30}\}$ (for the adjoint equations) are specified
6. the inequality conditions in Eq. (31) and (38) (not expanded for the sake of brevity) and the conditions at injection (Eq. (21)) are evaluated.

In summary, letting $f_4 = f_3 + \Delta f$, the problem reduces to the determination of seven unknown parameters, $\{\alpha_1, \alpha_2, \alpha_3, \lambda_{10}, \lambda_{20}, \Delta f, t_f\}$, that lead the dynamical system to satisfying the three final conditions (21), which corresponds to minimizing to 0 the following auxiliary objective function:

$$J' = |r_d(t_f) - r_d| + \left| v_{I,d}(t_f) - \sqrt{\frac{\mu_E}{r_d}} \right| + |\gamma_{I,d}(t_f) - \gamma_d| \tag{39}$$

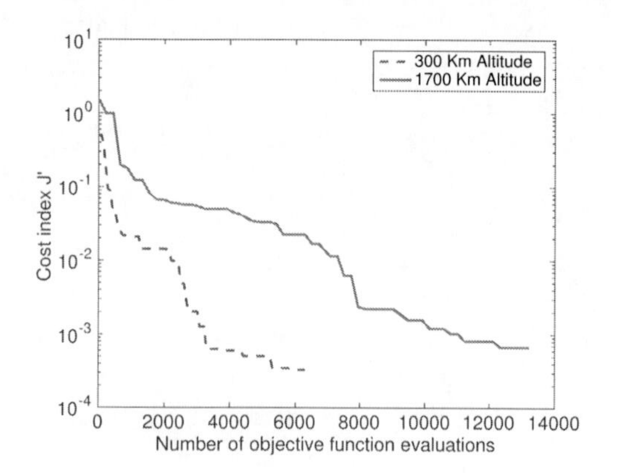

Figure 5. Error in the orbit injection evolution as a function of the number of objective function evaluations.

Minimizing J' to 0 implies satisfying the boundary conditions, while the minimization of the original objective function is ensured through the satisfaction of all the analytical conditions.

4.3. Numerical Results

The ascending trajectories are determined by employing canonical units: the distance unit (DU) is the Earth radius, whereas the time unit (TU) is such that $\mu_E = 1$ DU^3/TU^2. Thus DU=6378.165 km and TU=806.8 sec. The firework optimizer is used to find the unknown parameters. According to the description given in 4.3, there are different quantities to set: namely the number of fireworks N_F, the maximum number of sparks for each firework $N_S^{(max)}$ and the maximum number of objective function evaluations N_{evals}. The values used for the optimization in this work after some test runs are $N_F = 10$, $N_S^{(max)} = 30$ and $N_{evals} = 50000$. It is straightforward to demonstrate that increasing the maximum amount of sparks for a firework will improve the accuracy of the algorithm but in the end the general rule is to find a reasonable tradeoff between computation runtime and fitness reduction per iteration. Fig. 5 shows the objective evolution as a function of the number of evaluations. The computation ended when J' reached the order of 10^{-4}.

The initial condition for Eqs. (8)-(11) are listed in the Eqs. (40)-(41).

$$r(0) = R_E \;\; \gamma(0) = 86 \text{ deg} \tag{40}$$

$$v(0) = 1 \text{ m/sec} \;\; m(0) = m_0^{(1)} \tag{41}$$

where R_E (=6378.136 km) is the Earth radius.

To limit the solicitations due to dynamic pressure in the first three stages, the optimal values of the unknown angles of attack are sought in the range $0 \leq \alpha_1 \leq 5$ deg and $0 \leq \alpha_i \leq 10$ deg (i=2,3) while for the last stage $-1 \leq \lambda_{k0} \leq 1$ (k=1,2,3), $0 \leq \Delta f \leq \pi$, and 1 sec $\leq t_f \leq$ 30 sec. It is worth remarking that ignorability of the transversality condition allows defining arbitrarily the search space for the initial value of the Lagrange multipliers. This means that they can be sought in the interval $-1 \leq \lambda_{k0} \leq 1$ by the firework algorithm and only a posteriori their correct values (that also fulfill the transversality condition in Eq. (37)) can be recovered. Two test cases have been considered: R_d=R_E+300 km and R_d=R_E+1700 km. The main optimization results are reported in the Table 7:

Table 7. Optimal set of parameters

Altitude Km	300	1700
α_1 deg	3.65	2.59
α_2 deg	8.86	4.76
α_3 deg	9.85	4.93
λ_{01}	-0.17958	-0.13765
λ_{02}	-0.41587	-0.81500
Δf deg	15.33	29.06
t_f sec	24.4	22.8

Hence the coast arc durations Δt_{CO} are 295.2 sec and 839.4 sec, respectively for the 300 and 1700 km cases. Fig. 7 and 8 portray the state components (radius, velocity, flight path angle and mass) for each subrocket obtained with the optimized parameters in Table 7, while Fig.6 shows the optimal thrust pointing angle for the last stage, where a near horizontal burn circularizes the orbit at the desired altitude.

From inspecting the results summarized in Fig. 7-8 it is apparent that the final conditions at injection are fulfilled in a satisfactory way. The final payload masses obtained through the optimization are respectively 223.87 kg and

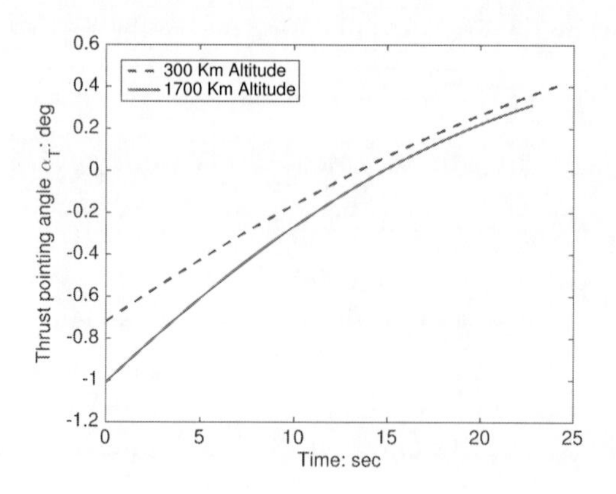

Figure 6. Control law for the last stage.

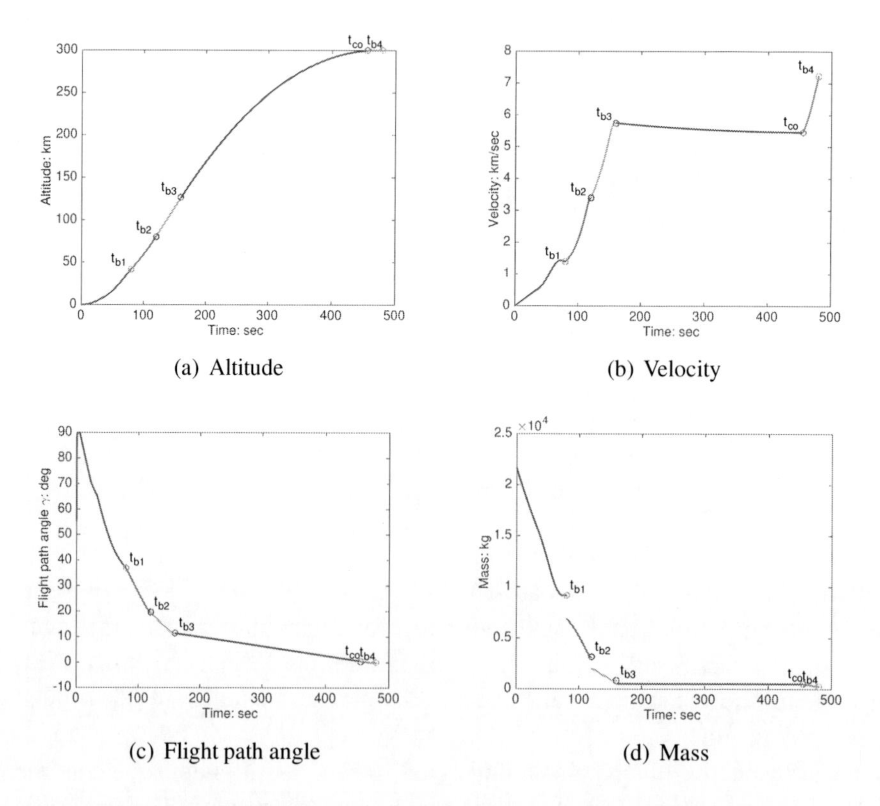

(a) Altitude

(b) Velocity

(c) Flight path angle

(d) Mass

Figure 7. State variables for all the four stages, 300 km case.

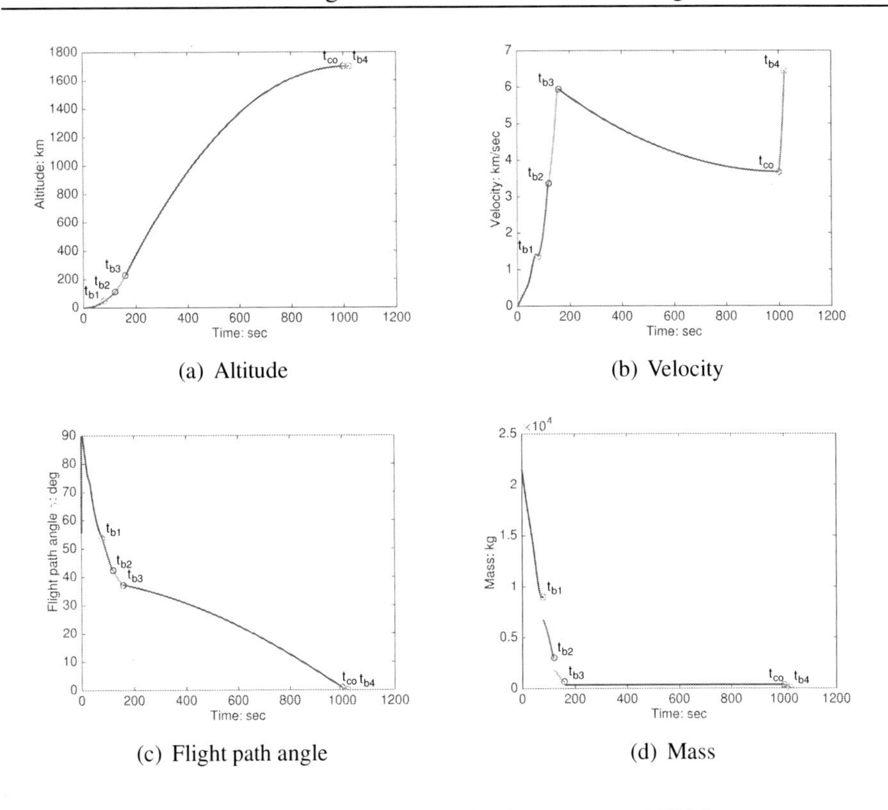

(a) Altitude

(b) Velocity

(c) Flight path angle

(d) Mass

Figure 8. State variables for all the four stages, 1700 km case.

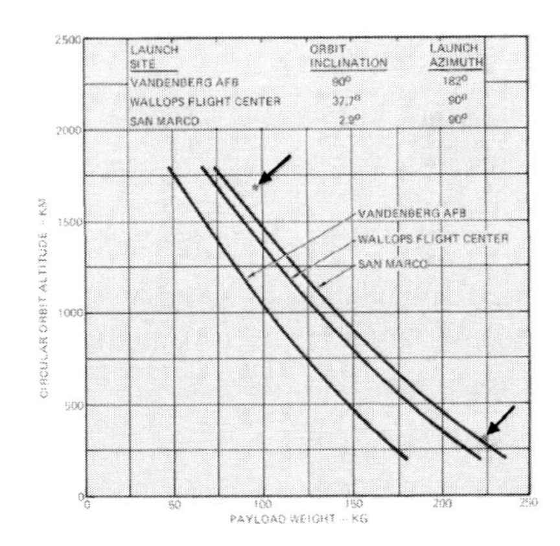

Figure 9. Circular orbit performances from the Scout manual [18], the $\star$ indicates the results of the optimization.

89.98 kg, obtained subtracting the structural mass (38.33 kg) from the final mass. These results outperform the ones presented in Fig. 9, taken from the Scout manual ([18]).

Conclusion

The generation of an optimal trajectory for multistage launch vehicles is a challenging problem, treated with different approaches in the past. This work proposes and successfully applies a simple technique for generating near-optimal two-dimensional ascending trajectories for multistage rockets, for the purpose of performance evaluation. A simple implementation of firework algorithm is employed, in conjunction with the analytical necessary conditions for optimality, applied to the upper stage trajectory. With regard to the problem at hand, the unknown parameters are (i) the aerodynamic angles of attack of the first three stages, (ii) the coast time interval, (iii) the initial values of the adjoint variables conjugate to the upper stage dynamics, and (iv) the thrust duration of the upper stage. The numerical results unequivocally prove that the methodology at hand is rather efficient, effective, and accurate, and definitely allows evaluating the performance attainable from multistage launch vehicles with very realistic aerodynamic and propulsive modeling. Specifically, the method at hand leads to results that outperform those already reported for the Scout vehicle.

Appendix

Firework Algorithm

The fireworks algorithm is a novel swarm intelligence method for parameter optimization problems [17]. It is a technique inspired by the firework explosions in the night sky. The concept that underlies this methodology is relatively simple: a firework explodes in a point of a n-dimensional space, with an amplitude and number of sparks that are determined dynamically through evaluation of the objective function at that point. The succeeding generation is chosen by including the best sparks (in relation with the objective function), and these become the new fireworks. This process is iterated until the termination criterion is met. For instance, this can be reaching the maximum number of function evaluations or reaching the desired fitness. If χ represents the parameter set, as a starting point N_F fireworks are generated randomly in the feasible space,

$$a \leq \chi \leq b \tag{42}$$

At iteration j, each firework and spark, denoted respectively with the index i and l, is associated with a particular determination of the parameter set, corresponding to a solution of the problem, with a specific value of the objective function,

$$\chi^{(j)}(i) = [\chi_1^{(j)}(i), \ldots, \chi_n^{(j)}(i)], \;\; j = 1, \ldots, N_F \tag{43}$$

$$\chi^{(j)}(l) = [\chi_1^{(j)}(l), \ldots, \chi_n^{(j)}(l)], \;\; l = 1, \ldots, N_{S,tot} \tag{44}$$

The actual number of sparks, $N_{S,tot}$, varies dynamically iteration after iteration, never exceeding the maximal value $2N_F N_S^{(max)}$, where $N_S^{(max)}$ represents the maximum number of sparks for each firework. In greater detail, the following steps compose the generic iteration j:

1. update the minimal spark amplitude, $A_{min}^{(j)}$

$$A_{min}^{(j)} = A_i - \frac{A_i - A_f}{N_{EV}^{(max)}} \sqrt{N_{EV}^{(j)} \left(2N_{EV}^{(max)} - N_{EV}^{(j)} \right)} \tag{45}$$

   where $N_{EV}^{(max)}$ is the maximum number of objective function evaluations, $N_{EV}^{(j)}$ is the number of evaluations performed prior to iteration j, whereas A_i and A_f are two reference minimum amplitudes, with $A_i > A_f$;

2. for i=1, ..., N_F: evaluate the objective function associated with firework i, $J^{(j)}(i)$;

3. in relation to the objective function, determine the global best and worst parameter set, $\mathbf{Y}^{(j)}$ and $\mathbf{Z}^{(j)}$, associated with the current fireworks,

$$i_m = \arg\min_i J^{(j)}(i) \;\; i_M = \arg\max_i J^{(j)}(i) \tag{46}$$

   which implies

$$\mathbf{Y}^{(j)} = J^{(j)}(i_m) \tag{47}$$

$$\mathbf{Z}^{(j)} = J^{(j)}(i_M) \tag{48}$$

4. for i=1, ..., N_F: calculate the number of sparks

$$N_S(i) = \text{round}\left\{ N_S^{(max)} \frac{J^{(j)}(i_M) - J^{(j)}(i) + \epsilon}{\sum_{r=1}^{N_F} [J^{(j)}(i_M) - J^{(j)}(i)] + \epsilon} \right\} \tag{49}$$

where ϵ is a tiny constant, whereas the operator round selects the integer part of the expression in parenthesis; if $N_S(i) = 0$, then $N_S(i)$ is set to 1;

for each component of χ (associated with k), calculate the explosion amplitude

$$A_k(i) = A_k^{(max)} \frac{J^{(j)}(i) - J^{(j)}(i_m) + \epsilon}{\sum_{r=1}^{N_F} [J^{(j)}(i) - J^{(j)}(i_m)] + \epsilon} \tag{50}$$

where $A_k^{(max)}$ is the maximal allowed amplitude, typically the search space amplitude $(b_k - a_k)$;

for each component of $\chi(i)$, generate two random numbers with uniform distribution (with limiting values indicated in parenthesis), $r_1(0, 1)$ and $r_2(-1, 1)$. Then

$$\begin{cases} \text{if } r_1 \leq 0.5 \rightarrow \chi_k^{(j)}(l) = \chi_k^{(j)}(i) + A_k(i) r_2 \\ \text{if } r_1 > 0.5 \rightarrow \chi_k^{(j)}(l) = \chi_k^{(j)}(i) \end{cases} \tag{51}$$

with l=1, ..., $N_S(i)$; for each component of $\chi(l)$ (l=1, ..., $N_S(i)$), generate a random number with Gaussian distribution (zero average value, standard deviation equal to 1), $n_1(0, 1)$, and a random number with uniform distribution, $r_3(0, 1)$. Then

$$\begin{cases} \text{if } r_3 \leq 0.5 \rightarrow \chi_k^{(j)}(N_s(i) + l) = \chi_k^{(j)}(l) \\ \quad + n_1[Y_k^{(j)} - \chi_k^{(j)}(l)] \\ \text{if } r_3 > 0.5 \rightarrow \chi_k^{(j)}(N_s(i) + l) = \chi_k^{(j)}(l) \end{cases} \tag{52}$$

with l=1, ..., $N_S(i)$; for each component of $\chi(l)$ (l=1, ..., $2N_S(i)$),

$$\begin{aligned} &if\ \chi_k^{(j)}(l) < a_k\ or\ \chi_k^{(j)}(l) > b_k \\ &\text{then } \chi_k^{(j)}(l) = a_k + r_4(0, 1)(b_k - a_k) \end{aligned} \tag{53}$$

5. the fireworks and the sparks are ordered in relation to their respective objective function value. Then, N_R fireworks for the new generations are selected among the best elements in $\{\chi(i), \chi(l)\}_{i=1,\ldots,N_F;l=1,\ldots,2N_S(i)}$ whereas the remaining $(N_F - N_R)$ fireworks are selected randomly among the remaining elements.

The *explosion operator* associates a reduced number of sparks to the fireworks that have higher values of the objective function to minimize, thus improving local search in the proximity of the most promising solutions. The *explosion amplitude* depends also on the objective function associated with each firework, and assumes greater values as the objective decreases. However, excessive contraction of the amplitude is avoided through adoption of the minimal amplitude , which is also dynamically updated, without reaching 0. The *displacement operator* creates new sparks according to the amplitude calculated at the previous step. The *Gaussian operator* introduces a displacement with random distribution toward the best position yet located up to the current iteration. The *mapping operator* avoids violation of the parameter bounds . Finally, the *elitism-random selection mechanism* preserves the best sparks and fireworks in the generation to come, while introducing new fireworks through random selection, in order to avoid premature convergence. In the end, the firework algorithm has several features that ensure satisfactory performance in parameter optimization problems, because both local search and global search are effectively performed through combination of the various operators described in this appendix.

References

[1] D. J. Bayley, R. J. H. Jr., J. E. Burkhalter, and R. M. Jenkins. Design optimization of a space launch vehicle using a genetic algorithm. *Journal of Spacecraft and Rockets*, 45(4):733–740, 2008.

[2] W. B. Blake. *Missile DATCOM User's Manual – 1997 Fortran 90 Revision*. United States Air Force, 1998.

[3] A. J. Calise, S. Tandon, D. H. Young, and S. Kim. Further improvements to a hybrid method for launch vehicle ascent trajectory optimization. In *AIAA Guidance, Navigation, and Control Conference and Exhibit*, Denver, CO, 2000.

[4] P. F. Gath and A. J. Calise. Optimization of launch vehicle ascent trajectories with path constraints and coast arcs. *Journal of Guidance, Control and Dynamics*, 24(2):296–304, 2001.

[5] R. Jamilnia and A. Naghash. Simultaneous optimization of staging and trajectory of launch vehicles using two different approaches. *Aerospace Science and Technology*, 2011.

[6] P. Lu, B. J. Griffin, G. A. Dukeman, and F. R. Chavez. Rapid optimal multiburn ascent planning and guidance. *Journal of Guidance, Control and Dynamics*, 31(6):45–52, 2008.

[7] P. Lu and B. Pan. Trajectory optimization and guidance for an advanced launch system. In *30th Aerospace Sciences Meeting and Exhibit*, Reno,NV, 1992.

[8] L. Mangiacasale. Meccanica del volo atmosferico. Terne di riferimento, equazioni di moto, linearizzazione, stabilità. [Mechanism of atmospheric flight. Backhoes, motion equations, linearization, stability.] *Ingegneria 2000*, 2008.

[9] P. Martinon, F. Bonnans, J. Laurent-Varin, and E. Trélat. Numerical study of optimal trajectories with singular arcs for an ariane 5 launcher. *Journal of Guidance, Control and Dynamics*, 32(1):51–55, 2009.

[10] A. Miele. Multiple-subarc gradient-restoration algorithm, part 1: Algorithm structure. *Journal of Optimization Theory and Applications*, 116(1):1–17, 2003.

[11] A. Miele. Multiple-subarc gradient-restoration algorithm, part 1: Application to a multistage launch vehicle design. *Journal of Optimization Theory and Applications*, 116(1):19–39, 2003.

[12] M. Pontani. Particle swarm optimization of ascent trajectories of multistage launch vehicles. *Acta Astronautica*, 2014.

[13] M. Pontani and P. Teofilatto. Simple method for performance evaluation of multistage rockets. *Acta Astronautica*, 94:434–445, 2014.

[14] J. E. Prussing and B. A. Conway. *Orbital Mechanics*. Oxford University Press, New York, NY, 1993.

[15] M. D. Qazi, H. Linshu, and T. Elhabian. Rapid trajectory optimization using computational intelligence for guidance and conceptual design of multistage launch vehicles. In *AIAA Guidance, Navigation, and Control Conference and Exhibit*, San Francisco, CA, 2005.

[16] W. Roh and Y. Kim. Trajectory optimization for a multi-stage launch vehicle using time finite element and direct collocation methods. *Engineering Optimization*, 34(1):15–32, 2002.

[17] Y. Tan. *Fireworks Algorithm*. Springer, 2015.

[18] VoughtCorporation. *SCOUT Planning Guide*. Vought Corporation, 10 1976.

[19] N. Weigel and K. H. Well. Dual payload ascent trajectory optimization with a splash-down constraint. *Journal of Guidance, Control and Dynamics*, 23(1):45–52, 2000.

In: Advances in Mathematics Research
Editor: Albert R. Baswell
ISBN: 978-1-53612-512-2

Chapter 6

TIES AND REDUCTIONS FOR SOME SCHEDULING AND ROUTING PROBLEMS

Nodari Vakhania*
Centro de Investigación en Ciencias, UAEMor
Cuernavaca, Mexico

Abstract

Scheduling and transportation problems are important real-life problems having a wide range of applications in production process, computer systems and routing optimization when the goods are to be distributed to the customers using scarce available resources. In this chapter, we illustrate ties and relationships among some of these optimization problems. We consider scheduling problem with release and due dates, batch scheduling and vehicle routing problems. As we will show here, although these problems seem to have a little in common, a closer look at their parametric and structural properties can give us more insight into the "hidden" ties among these problems that may lead to efficient solution methods.

Keywords: discrete optimization problem, decision problem, scheduling with release times and due dates, vehicle routing problem, subset sum and bin packing problems, polynomial-time reduction

*Corresponding Author Email: nodari@uaem.mx.

1. Introduction

In the theory of formal languages the *mapping reduction* is a key concept. A formal language A mapping reduces to another language B if there exists a computable function f such that for any sting $s \in A$, $s \in A \iff f(s) \in B$. In other words, the membership of any string to language A can be determined by function f, given that there is a procedure for the establishment of the membership of any string to language B. In this way, language A can be recognized/decided by a recognition/decision procedure for language B. Languages are not mere formal entities, they can represent real-life problems as a set of *strings* over some alphabet. A *combinatorial optimization* problem is a formal mathematically stated problem that represents some task (arisen mainly in real life). In these problems, a set of objects is given, each object having one or more parameters. A *feasible solution* arranges these objects in some fashion so that a given set of restrictions (which arise naturally in real-life) are satisfied. A given objective function needs to be minimized/maximized under these restrictions, i.e., a feasible solution that achieves the corresponding extremal value of the objective function, the so-called *optimal solution* needs to be determined. Every minimization (maximization, respectively) problem has its *decision* version in which one merely wishes to know if there exists a feasible solution to the problem with the value of the objective function no more (no less, respectively) than some given constant. Using *binary search*, a procedure for solving a decision problem can be transformed in polynomial time into a procedure that solves the corresponding optimization problem. We give an example of such transformation in the next Section 2.

The mapping reducibility can be applied for the decision problems since every decision problem can be represented as a formal language. In combinatorial optimization theory, the reducibility among different problems is an important concept. If the reduction is polynomial, i.e., if the above mentioned function f can be computed in polynomial in the length of the input time, then the reduction becomes even more attractive as a polynomial-time solution procedure for one problem can be used for the solution of another one, i.e., one problem reduces to another one in polynomial time.

Polynomial-time reduction is basic in the *complexity* hierarchy of the combinatorial optimization problems. In an important class of intractable NP-hard problems, one problem reduces polynomially to another one from the class. Two apparently different problems may, in fact admit quite similar solution methods due to the (polynomial) reducibility. In this chapter, we illustrate the similarity of some seemingly distinct combinatorial optimization problems. In the rest of this section, we first give a brief description of the type of problems we are dealing with in this chapter, and then we will give a short sketch of how it is organized.

The *scheduling* problems constitute an important class of the combinatorial optimization problems (for general guideline in scheduling theory see, for example, Pinedo [2]). A job shop problem in industry, scheduling of information and computational processes, trains, ships and airplanes services are some examples of practical problems which are studied in scheduling theory. There is a general notion of orders or requests which are to be assigned to the given resources. Requests and resources, respectively, are typically referred to as *jobs* (operations or *tasks*) and *machines* (or *processors*), respectively. A job is a part of the whole work to be done, a machine is the means for the performance of a job. A basic restriction is that any machine can handle at most one job at a time (in some scheduling problems, for instance, in batch scheduling problems, this restriction is somehow relaxed, as we will see later in Section 4). In scheduling problems, a basic parameter of every job j is its size or *processing time* p_j. In a *feasible schedule* every job j is assigned unique time interval(s) with the total length p_j on a processor. So a schedule arranges an order in which the jobs are handled by the machines. A general aim is to find a feasible schedule that minimizes or maximizes a given objective function. For instance, a common objective function for scheduling problems is the maximum job completion time commonly referred to as the *makespan* denoted by $C_{\max}$.

The *Vehicle Routing Problems* (VRP) proposed by Dantzig and Ramser in early 1959 arise in vast amount in practical circumstances when the goods are to be distributed to the customers using a limited number of vehicles (see e.g., [3]). In general, the transportation problems form a significant part of practical real-life problems formalized as mathematical combinatorial optimization problems, the majority of which are computationally intractable.

The chapter is organized as follows. In the next Section 2 we describe basic scheduling problems in which jobs with release times and due dates or delivery times are to be processed by a single machine or by a group of parallel machines. We illustrate that these apparently different problems are equivalent. We consider the decision versions of these problems and specify how a solution to a decision version can be used for the solution of the corresponding optimization version. In Section 3 we describe other optimization problems and show how they can be reduced to the scheduling problems from Section 2. We dedicate Section 3.1 to the ties between vehicle routing problem and the above scheduling problems. Section 4 is dedicated to batch scheduling problems. We describe three basic versions of the batch scheduling and suggest fast polynomial-time algorithms for their solution. We also observe the ties of these problems with subset sum and bin packing problems. In Section 5 we give final remarks.

2. Scheduling with Release Times and Due Dates

2.1. A Basic Scheduling Problem

In this section we describe three versions of a scheduling problem with release and due dates. We show that the first two versions are equivalent, whereas the solution procedure for the third version can be used for the solution of the first two versions. Then we consider a version of vehicle routing problem and show how the solution to the scheduling problem can be used for the solution of that vehicle routing problem.

Our first scheduling problem is as follows. We are given n jobs $\mathcal{J} = \{1, 2, ..., n\}$ and m parallel *identical machines* (or *processors*) $\mathcal{M} = \{1, 2, ..., m\}$. Job $j \in \mathcal{J}$ becomes available only at its *release time* or *head* r_j, it needs continuous processing time p_j on any of the machines and also it needs an additional *tail* or *delivery time* q_j after the completion on the machine. The delivery is accomplished by an independent resource and hence needs no machine time in our model. The heads and tails are non-negative integers while the processing time p is a positive integer. A *feasible schedule* S assigns to each job j a starting time t_j^S and a machine from $\mathcal{M}$, such that $t_j^S \geq r_j$ and $t_j^S \geq t_i^S + p$, for any job i scheduled earlier on the same machine (the first inequality ensures that a job cannot be started before its release time, and the second reflects the restriction that each machine can handle only one job at any time). The *completion time* of job j in S is $c_j^S = t_j^S + p$ and the *full completion time* of j in S is $C_j^S = c_j^S + q_j$. Our objective is to find an *optimal schedule*, i.e., a feasible schedule S that minimizes the maximal full job completion time $|S| = \max_j C_j$, called the *makespan*.

Let us now consider a seemingly similar scheduling problem, though formulated in a different way. In that problem, the jobs have no tail, instead, every job j has *due date* d_j, which is the desirable time by which this job can be finished. The *lateness* of job j in schedule S, $L_j^S = c_j^S - d_j$. Now we wish to minimize the maximal job lateness $L_{max} = \max_{j \in \mathcal{J}} L_j$ instead of the maximal job completion time as in the first described version.

Observation 1. *The two scheduling problems as defined above are equivalent.*

Proof. Given an instance of the version with tails, we construct an equivalent instance of the version with due dates as follows. Let K be a large constant, say a number greater than the maximum job tail. Then for the instance with due dates, the due date of every job $j \in \mathcal{J}$ is defined as $d_j = K - q_j$. Note that the difference between tails and the newly derived due dates is a constant (to larger tails smaller due dates correspond, which is what one would naturally expect – the larger is job tail the more urgent is the job, the smaller is job due date, the more urgent is that job). Let S be an optimal solution to the problem with tails and S' be the corresponding schedule with due dates. Since the due dates of the jobs remain unchanged, $c_j^S = c_j^{S'}$, for all $j \in \mathcal{J}$. Furthermore, by the way we have defined job due dates, job o realizing the maximum full completion time $\mathcal{C}_o^S$ in schedule S realizes the maximum job lateness L_o in schedule S'. But since schedule S is optimal, both values $\mathcal{C}_j^S$ and L_o are the minimal possible. Hence, S' is an optimal solution.

In the other direction, given an instance with due dates, we transform it to an equivalent instance with tails. We now take a constant K larger than the maximum job due date $d_{max} = \max j \in \mathcal{J} d_j$ and define the tail of every job $j \in \mathcal{J}, q_j = K - d_j$. Let schedules S and S' be defined as above. Similarly, job o realizing the maximum lateness L_o in schedule S' will realize the maximum full completion time $\mathcal{C}_o^S$ in schedule S and the optimality of schedule S follows.

□

2.2. The Decision Version of the Scheduling Problem

In the introduction we have mentioned about optimization and decision versions of the same problem. We now define a decision version of the above scheduling problem with due dates. In the decision version, job data are defined as in the latter version with due dates. Our objective now is to verify whether there exists a feasible schedule in which the lateness of no job is more than a constant K.

Next, we specify how a solution procedure for this decision problem can be used for the minimization of $L_{\max}$. We use binary search, for which we first need to determine lower and upper bounds for the possible maximum job lateness.

The lower bound l on the maximum job lateness can be determined as

$$l = \min_{j \in \mathcal{J}} p_j - d_j.$$

Indeed, note that the maximum lateness will be a negative number if all jobs complete before their due dates. But no job j can be completed before time p_j, hence $p_j - d_j$ is a lower bound on the maximum job lateness.

We can obtain an upper bound u on the maximum job lateness by creating a feasible schedule to a given scheduling instance. Then the maximum job lateness in that schedule will obviously be the desired upper bound u. To create a feasible solution, we can apply some heuristic algorithm. Below we describe the so-called EDD (Earliest Due Date) heuristics commonly used for the solution of our scheduling problem.

EDD-heuristic creates n scheduling times so that at each of these times a new job is assigned. For the simplicity of presentation, we first describe the heuristic for a single machine case. Initially, the earliest scheduling time is the minimum job release time. Iteratively, among all jobs released by a given scheduling time, a job with the minimum due date is scheduled on the machine ties being broken by selecting any longest available job. Once a job completes on the machine, the next scheduling time is set to the maximum between the completion time of that job and the minimum release time of a yet unscheduled job. Note that since the heuristic always schedules an earliest released job every time the machine becomes idle, it creates no avoidable gap. Since at every scheduling time the search for a maximal element in an ordered list is carried out and there are n scheduling times, the time complexity of the heuristic is $O(n \log n)$.

If we have m parallel identical machines instead of a single one, a ready job with the smallest due date is repeatedly determined and is scheduled on the machine with the minimal completion time (ties being broken by selecting the machine with the minimal index.

Observation 2. *The optimization version of the scheduling problem can be solved by the application of a procedure for the solution of the decision version* $O\log(u - l)$ *times.*

Proof. Having defined the lower and upper bounds as described above, we carry out the binary search within the interval $[l, u]$ invoking a solution procedure $\mathcal{P}$ for the decision version for the trial values within the interval $[l, u]$. For every positive answer of procedure $\mathcal{P}$ the next trial value is smaller than the last tried one, and for every negative answer of procedure $\mathcal{P}$ the next trial value is larger than the last tried one. Since the total number of calls of procedure $\mathcal{P}$ is $O\log(u-l)$, the total cost for the solution of the optimization version is $O\log(u-l)$ multiplied by the time complexity of procedure $\mathcal{P}$. □

3. The Scheduling Problem vs Other Optimization Problems

Although our scheduling problem is NP-hard, there are a number of efficiently solvable cases and approximation algorithms for it. For example, for a single machine, EDD-heuristic gives a 2-approximation solution, i.e., a solution, at most twice worse than an optimal one, whereas, in practice, its performance is much better [7]. If there are two allowable processing times p and $2p$ then the problem is polynomially solvable [5] and it remains such if the maximal job processing time and the differences between job release times are bounded by a constant [6]; other optimality conditions were recently studied in [1]. With m machines, the problem can still be solved in polynomial time if all jobs have the same length [4].

The problem has numerous immediate real-life applications. For instance, consider a car production process where cars are assembled from different parts which are produced and delivered to the factory independently. These parts need to be further processed during the whole assembly process and their release times are set to the dates when they arrive at the factory. It is desirable to process every part by some due date; if this is not possible, the manufacturer wishes to minimize the maximum lateness during the processing of any part.

As another example, consider CPU time sharing in operating systems, the jobs being the processes that are to be handled by the processor(s). The processes which arrive over time at their release times need some estimated time on the processor and have different priorities, which are reflected by their due dates. The objective is again to minimize the maximum deviation from the desired serving times, i.e., to minimize the maximum lateness.

In a version of the interval distribution problem, a finite set of intervals on the real axes are given, and the goal is to re-distribute them so that no two intervals intersect (a restriction defining a feasible solution) and the maximum interval shifting value is minimized. This problem naturally arises in wireless sensor network transmission range distribution where we wish to distribute the intervals so that they do not intersect (to avoid unwanted disturbances) and the resultant distribution is as close to the initial one as possible.

This problem can be seen as a special case of the single-machine version of our scheduling problem. We associate intervals with jobs the length of every job being the length of the interval, associated with that job; the given intervals on the time exes are defined as follows: the left and right endpoints of every interval are associated with the release time and due date of the corresponding job. Then finding a schedule with minimal maximum lateness will give us a solution to the interval problem with the minimal interval shifting distance.

3.1. Vehicle Routing Problem with Time Windows

One of the most practical and also complex combinatorial optimization problems is the earlier mentioned Vehicle Routing Problem. We first state the basic (uncapacitated) version of this problem and then describe a particular version of it with so-called time windows which can be solved using a solution method to our scheduling problem, as we show later.

We are given an undirected weighted (complete) graph $G = (V, E)$ with edge weights w_e, for all $e \in E$, a distinguished node v_d from set V (called *depot*) and a positive integer number k. $(i, j) \in E$ is the edge connecting node i with node j. For any $Y \subseteq V$ containing node v_d, a *tour* T_Y defined by set Y is a directed cycle that starts at that node, visits every node in Y exactly once and returns to the same node v_d; in other words, $T_Y = (i_1, i_2, \ldots, i_l, 1)$, where $(i_2, \ldots, \imath_l)$ is an enumeration of the nodes in set Y not including node v_d and $i_1 = v_d$. The *cost* of tour T_Y, $c(T_Y)$ is the sum of the weights of the edges on this cycle, i.e., $c(T_Y) = w(i_1, i_2) + w(i_2, i_3) + \cdots + w(i_l, 1)$. VRP aims to find the partition of nodes from set $V \setminus \{v_d\}$ in k subsets $V_1, \ldots, V_k$ with the minimal possible total cost; that is, with the minimal $\sum_{i=1,2,\ldots,k} c(V_i)$.

We may give a practical interpretation to the above described version of the vehicle routing problem assuming that k is the number of identical vehicles (one for each of the subsets V_i) that can travel among the n customers or cities, and there is one special location called depot. The distances between any pair of locations is known: the weight $w(i, j)$ is the distance between customers i and j. We assume that there are the goods that are concentrated in the depot. They need to be distributed from the depot to the customers using one or more (identical) vehicles. The restrictions on how this distribution should be done define the set of all feasible solutions to the problem. A valid vehicle tour is initiating at depot, visits some of the customers and returns to depot. All customers must be served, i.e., every customer is to be included in exactly one tour and the total travel distance is to be minimized.

Without time windows, a basic vehicle routing problem is a straightforward generalization of another well-known optimization problem, the so-called Traveling Salesman's Problem (TSP) in which one looks for a shortest tour visiting all the given n cities exactly once. In the Euclidean TSP the distances between the cities are Euclidean distances, in more general setting, they are arbitrary unrelated (positive) numbers. The vehicle routing problem with a single vehicle, i.e., with $k = 1$ is TSP. Multiple TSP, a generalization of TSP with k-tours, k-TSP, is a VRP with k vehicles. Euclidean 1-TSP is already intractable NP-hard problem.

The version of the vehicle routing problem that we consider here is one with time windows. In this setting, every customer can only be served within a certain time window (interval), and there is also a valid time interval given for the depot. The objective is to find the minimal number of vehicles for which there exists a feasible solution satisfying the restrictions. Then we wish to minimize the total service/travel time/distance of the determined number of vehicles.

Observation 3. *A feasible solution for the basic vehicle routing problem can be found in time $O(n)$.*

Proof. We basically choose some order to visit the customers for each of the k vehicles so that no customer is left outside of one of these tours partitioning first the set of nodes into k disjoint sets. This can be done in linear time by simply choosing an approximately same number of jobs in each subset (rounding down the magnitude n/k and then adding the remained nodes to one of the subsets). Then, for every created subset, we generate a tour in time, linear in the number of elements in each subset by merely visiting the nodes in any order. In this way, a feasible solution will be created in time $O(n) + O(n) = O(n)$. □

In the setting with time windows, in opposite to the basic case, finding a feasible solution is already a difficult task; in fact, there may exist no feasible solution. If there exists a feasible solution, a common objective for the version with time windows is to minimize the number of the used vehicles, i.e., find the minimal number k such that there exists a feasible solution to the problem with k vehicles. Then, as in the basic version, we construct the k tours with the objective to minimize the total service time.

Proposition 1. *The problem of finding a feasible solution to the vehicle routing problem with time windows can be reduced to the problem of finding a feasible solution of the decision version of the scheduling problem in polynomial $O(n)$ time, where n is the number of nodes in the vehicle routing instance.*

Proof. Our proof is constructive. We describe how an algorithm to our scheduling problem can be used to find a feasible solution to the vehicle routing problem with time windows.

First, we specify the general reduction scheme, i.e., the correspondence between the objects and the parameters of the two optimization problems, that takes time $O(n)$. We associate with every customer (and the depot) a unique job, and with the time window of each customer the time interval determined by the release time and the due date of the job corresponding to that customer. Then note that an instance of our scheduling problem *with a single machine* is already determined.

Now we consider the decision version for the created single-machine instance in which we set the available lateness to 0; in other words, this decision version asks for a feasible solution to the scheduling problem in which all the jobs complete by (at or before) their due dates (if there exists such a schedule then an algorithm that solves this problem gives a "yes" answer, otherwise, it gives a "no" answer).

It straightforwardly follows from this reduction that if the answer to the above decision version of the scheduling problem is "no" (i.e., there exists no feasible solution in which all jobs complete by their due dates), then there exists no feasible solution to the corresponding vehicle routing problem with a single vehicle. Therefore, if the above decision problem has no feasible solution, the feasible solution to the vehicle routing problem may only exist for two vehicles, i.e., for $k = 2$. Hence, we may restrict our attention to the decision problem with two identical machines and similarly apply our algorithm to that two-machine scheduling problem. Again, for the positive answer we halt, and now for the negative answer we consider the scheduling problem with three parallel machines. We continue in this fashion until we find the minimal number of machines for which the corresponding scheduling instance has a feasible solution.

At this stage, note that if the decision problem with $k = n$ machines has no feasible solution then there exists no feasible solution for the vehicle routing problem for any number of vehicles, hence our procedure may halt. It follows that the reduction procedure will need to create at most n decision instances of the scheduling problem to determine a feasible solution to the vehicle routing problem with the minimal possible number of vehicles or to establish that there exists no feasible solution to the given instance of the vehicle routing problem with time windows. This number can, in fact, be reduced to $O(\log n)$ if we apply binary search within the interval $[1, n]$. □

4. Batch Scheduling

In this section we consider another related problem, a *batch scheduling* problem. Unlike traditional scheduling machine, a *batch machine* may handle two or more jobs simultaneously. From the practical point of view, a batch machine can also be seen as a transportation vehicle that can ship a number of completed products to the customer(s) simultaneously. The jobs here can be seen as the finished products that need to be delivered to the customers.

In all versions of a basic batch scheduling problem that we consider, n jobs and one or more batch machines are given. Job j gets ready at time r_j and it is characterized by its *size* p_j that is the required processing or delivery time for job j. Once job j gets ready, it can be processed (or delivered to the customer) by a batch machine. Batch machine B has a limited *capacity* $C(B)$ (a positive natural number). In different versions in batch scheduling, the capacity $C(B)$ may serve for different purposes: in some versions it is an upper limit on the total *number* of jobs that machine B may handle, and in other versions it is the total *size* of the jobs that machine B may handle simultaneously. In both cases, the maximum job size is the time that machine B will take to process/deliver all the jobs assigned to that machine.

In the next subsections we define different versions of the batch scheduling problem which vary in the machine environment and the objective function.

4.1. Restricting the Number of Jobs

In this subsection we consider batch capacity as the restriction on the total number of jobs that might be assigned to a single batch.

Batch Scheduling Problem 1 (BSP-1). We have the unlimited number of batch machines with the capacity C that restricts the total number of jobs assigned to it. The size of a job is the time that takes its delivery to the customer. The delivery time of a batch is the maximum job size in that batch. Our objective is to minimize the total delivery time of all the jobs/batches. In the off-line version abbreviated BSP-1(off-line), the n jobs with their release and delivery times are given in advance, whereas in the on-line version BSP-1(on-line) a job data becomes known only when that job gets released, and the total number of jobs is not known in advance.

For the off-line case we propose the following optimal strategy for scheduling the n jobs into the batches, **Algorithm-[BSP-1-off-line]**:

Step 1. Sort the jobs in non-increasing order of their delivery times.

Step 2. While there are unassigned jobs, select the next C jobs from the list and schedule their delivery as a single batch starting at the maximum job release time in that batch; if there remain less than $C(B)$ unassigned jobs in the list, then assign them all to machine B and deliver this as a single batch at the maximum job release time in that batch (note that, in total, $\lceil n/k \rceil$th batches are created).

Theorem 1. *Algorithm-[BSP-1-off-line] returns an optimal solution to BSP-1(off-line) in time* $O(n \log n)$.

Proof. First we observe that the partition of Algorithm-[BSP-1-off-line] yields a feasible solution since each batch is delivered at the maximum job release time in that batch. Let π_i be the maximum job delivery time in batch i, $i = 1, \ldots, \lceil n/k \rceil$. Then $\sum_{i=1,\ldots,\lceil n/k \rceil} \pi_i$ is the total delivery time of all the jobs in the solution created by Algorithm-[BSP-1-off-line]. This magnitude is easily seen to be the minimal possible using a simple interchange argument: by swapping two elements in any two batches, the delivery time of any of these batches may only be increased, as the interchange may only increase the maximum job delivery time in one of the rearranged batches.

As to the time complexity, the initial sorting at Step 1 takes time $O(n \log n)$. During the formation of each batch, we spend linear time to group the jobs. While adding the jobs to a formed group, we keep the track of the maximum job release time with no extra cost. □

Now an on-line algorithm for BSP-1(on-line), **Algorithm-[BSP-1-on-line]**, repeatedly, waits until time $t = r_j$ when Cth yet unassigned job j gets released, and assigns all these C unassigned released jobs to a single batch and delivers it at time t.

We may easily observe that Algorithm-[BSP-1-on-line] not necessarily returns an optimal solution. For instance, let the batch capacity $C = 2$, jobs with delivery times 10 and 1 be released at time 0, and jobs with the same delivery times 10 and 1 be released at time 2. Algorithm-[BSP-1-on-line] will assign the first two jobs to one batch delivered at time 0 with delivery time 10, and the last two jobs to the second batch delivered at time 2 with the same delivery time 10. But there exists another solution, in which the two jobs with delivery time 10 are assigned to one batch, and the other two jobs with delivery time 1 are assigned to the other batch, both delivered at time 2. The delivery of the first batch takes time 10 and that of the second one takes time 1. Hence, the total delivery time in the second solution is 11 whereas the first solution gives the total delivery time equal to 20.

At the same time, if all jobs have the same processing time, the above on-line algorithm Algorithm-[BSP-1-on-line] will always create an optimal solution:

Theorem 2. *Algorithm-[BSP-1-on-line] finds an optimal solution to problem BSP-1(on-line) if all jobs have the same processing time.*

Proof. First note that the maximum job delivery time in any formed batch of any feasible solution is the same since all jobs have the same size. Because of the same reason, the total number of the batches formed by Algorithm-[BSP-1-on-line] is the minimal possible. These two observations together obviously prove the theorem. □

4.2. Restricting the Total Size of Jobs in One Batch

In this subsection we consider batch capacity as the restriction on the total size of jobs that might be assigned to a single batch.

Batch Scheduling Problem 2 (BSP-2). We have unlimited number of batch machines with the capacity C, that, this time, restricts the total size of jobs assigned to it. The size of a job is the time that takes a batch machine to process that job. The processing time of a batch is the maximum job size in that batch. Our objective is to minimize the total processing time of all the batches. Similarly as in the previous subsection's problem, in the off-line version, abbreviated BSP-2(off-line), the n jobs with their release and processing times are given in advance, whereas in the on-line version BSP-2(on-line) a job data becomes known only when that job gets released, and the total number of jobs is not known in advance.

Batch Scheduling Problem 3 (BSP-3) we define as BSP-2 with the only difference in the objective function: now we wish to minimize the number of the formed batches.

Observation 4. *An optimal solution to problem BSP-2 (off-line) is not necessarily an optimal solution to BSP-3 even if all jobs are released simultaneously. In other words, the number of the formed batches containing a solution that minimizes the total processing time of all batches not necessarily is the minimum possible (i.e., there may exist another feasible solution to the same instance with less number of the formed batches).*

Proof. We prove the observation by giving a problem instance for which an optimal solution to problem BSP-2 is different to that of problem BSP-3 explicitly. Indeed, consider a problem instance with 10 jobs, two long jobs with processing time 7, and 8 short jobs with processing time 2. Let the batch capacity be 15 and all jobs be released at time 0.

An optimal solution to problem BSP-3 forms only two identical batches, each of them containing one long job with processing time 7 and four short jobs with processing time 2. The total processing time of both batches is then 7+7=14. At the same time, an optimal solution to problem BSP-2 consists of three batches, the first one containing two long jobs with processing time 7, the second one containing seven short jobs with processing time 2, and the third batch contains the only remained short job with processing time 2. The total processing time of the three formed batches is 7+2+2=11. As we can see, the total processing time of the second solution with three batches is less than that of the first solution with two batches which proves our observation. □

4.2.1. Algorithms for Problems BSP-2 and BSP-3

In this subsection we propose heuristic algorithms for problems BSP-2 and BSP-3. For these problems, while forming a batch we must clearly take care of the total size of the jobs assigned to that batch. Although problems BSP-1, BSP-2 and BSP-3 are apparently similar, the latter two ones are essentially more complicated. Intuitively, an optimal use of the batch capacity becomes more difficult as it requires the solution of NP-hard *subset sum* and *bin packing* problems:

In subset sum problem, we are given n items, p_i is the size of item i, and number M (the capacity). The question is to find out if there exists some subset of the set of items such that the sum of the item sizes in that subset equals to M. In *bin packing* problem, n items with given sizes are to be packed in the minimal possible number of identical bins with the same capacity C.

The ties between problems BSP-2 and BSP-3 and the above two problems are obvious: an optimal solution to problems BSP-2 and BSP-3 needs to distribute the jobs into the batches (bins) using a given batch capacity in some optimal manner.

Our heuristic **Algorithm-[BSP-2-off-line]** for problem BSP-2 (off-line) intends to join the largest possible number of long jobs into a single batch aiming in this way to minimize the sum of the maximum job processing times over all batches. As in Algorithm-[BSP-1-off-line], at the preprocessing stage the jobs are sorted according to their processing times:

Step 1. Sort the jobs in non-increasing order of their processing times.

Step 2. While there are unassigned jobs open a new batch: iteratively assign the next job from the list to that batch until the next selected job does not fit within the remained empty space of this batch; start the formed batch at the maximum job release time in that batch. Repeat Step 2.

Our heuristic algorithm **Algorithm-[BSP-3-off-line]** for problem BSP-3 (off-line) tries to use the batch capacity by leaving less idle space. The basic idea is to fill in all the batches uniformly longer jobs ahead shorter ones retaining in this way shorter jobs to the end. Since we do not know the required number of batches in advance, we take an estimated lower bound from the observation below.

Observation 5. *An optimal solution to problem BSP-3 (off-line) contains at least $\lfloor \sum_j p_j/C \rfloor$ different batches.*

Proof. For the best outcome, all the batches are completely filled in by the n jobs without any remained empty space in any batch. In this case, the total number of batches will be $\lfloor \sum_j p_j/C \rfloor$ which proves the observation. □

Algorithm-[BSP-2-off-line] initially creates $\lfloor \sum_j p_j/C \rfloor$ batches and assigns jobs to these batches by using First Fit Decreasing (FFD) strategy:

Step 1. Sort the jobs in non-increasing order of their processing times.

Step 2. Iteratively, assign the next job from the list to the next batch (from the initially created $\lfloor \sum_j p_j/C \rfloor$ batches) until no more job can fit into any of the batches. Start each batch at the maximum job release time in that batch.

Step 3. If an unassigned job is left, then assign them in the order as they appear in the list creating the necessary amount of the new batches.

Conclusion

We have described a number of combinatorial optimization problems and have illustrated ties between these apparently different problems. In general, such ties apply to many other classes of combinatorial optimization problems. In fact, the notion of NP-completeness also relies on similar relationships between different combinatorial optimization problems. In this chapter, we have concentrated our attention on some basic scheduling and vehicle routing problems. We have described fast heuristics for three batch scheduling problems that we abbreviated as BSP-1, BSP-2 and BSP-3. As we have seen, problem BSP-1 can be solved in polynomial time, whereas problems BSP-2 and BSP-3 turn out to be intractable. In these three batch scheduling problems with jobs released at different time moments, we have imposed no penalty for the late completion of the jobs that would essentially complicate them. For future work, it would be interesting to consider the extensions of these problems in which, as in our first scheduling problem, jobs are also characterized by their due dates and the penalty for the completion of jobs behind their due dates is imposed.

References

[1] E. Chinos and N. Vakhania. Adjusting scheduling model with release and due dates in production planning. *Cogent Engineering* 4(1), p. 1-23 (2017) DOI: 10.1080/23311916.2017.1321175.

[2] M. Pinedo. Scheduling: Theory, Algorithms, and Systems. *Springer* p.1-621 (2012) http://www.springer.com/la/book/9781489990433.

[3] Toth P., Vigo D. (editors) Vehicle routing problem *SIAM monographs on discrete mathematics and applications* 386 SIAM, Philidelphia (2002).

[4] N. Vakhania. A better algorithm for sequencing with release and delivery times on identical processors. *Journal of Algorithms* 48, p.273-293 (2003).

[5] N. Vakhania. Single-Machine Scheduling with Release Times and Tails. *Annals of Operations Research*, 129, p.253-271 (2004).

[6] N. Vakhania, F. Werner. Minimizing maximum lateness of jobs with naturally bounded job data on a single machine in polynomial time. *Theoretical Computer Science* 501, p. 72-81 (2013).

[7] N. Vakhania, Dante Perez, and Lester Carballo. Theoretical Expectation versus Practical Performance of Jackson's Heuristic. *Mathematical Problems in Engineering* volume 2015, ID 484671 http://dx.doi.org/10.1155/2015/484671 (2015).

In: Advances in Mathematics Research
Editor: Albert R. Baswell
ISBN: 978-1-53612-512-2

Chapter 7

A CONTINUOUS FOUNDATION FOR DIMENSION AND ANALYTIC GEOMETRY

N. L. Bushwick, PhD*
The Northeast Autism Center, Scranton, PA, US

Abstract

The construction presented here, like systems of Aristotelian continua presented elsewhere, is designed to establish a geometry without using points. However, it goes further in that the foundation consists of elements that are completely abstract, rather than line segments whose universe is a line. Furthermore, the result could be modified to represent elements and spaces of multiple dimensions.

Introduction

The concept of dimension presents several problems in traditional mathematics. Some involve the relationship of the mathematical system to the physical world, others are internal problems of the mathematical system itself. Since the physical world is three dimensional, or, if time is included, four dimensional, certain problems arise when a mathematical system involving elements of less than three dimensions is applied to it. Most obvious is the problem of dimensionless points, but there are also problems involving entities of one or two dimensions.

*Corresponding Author Email: nlbushwick@northeast-autism.org.

Related problems of one form or another have been recognised since ancient times, most famously in Zeno's paradoxes, and have been dealt with in various ways. For those who saw mathematics itself as an expression of the true nature of reality, these problems could be seen as proofs either that the physical world is really atomic or, on the contrary, that it is continuous. For others, they could be seen instead as revealing the insufficiency of mathematical systems as models of the physical world [Tannery, 1885].

As traditional systems evolved from their original Euclidian foundations, they came, via Descartes [Descartes, 1886], Bolzano [Bolzano, 1950] and Cantor [Cantor, 1955], to adopt the discrete approach. Line segments became infinite collections of points rather than continua. Thus the problem of composing a one-dimensional line from points lacking dimension was avoided.

Recently, mathematical systems have been constructed based on line segments, which provide a continuous alternative [Hellman and Shapiro, 2012; 2013; Linnebo, Shapiro and Hellman, 2016]. In these, the problem is avoided by eliminating points entirely. However, these, like traditional systems, begin with concepts derived from experience in the physical world, such as lines and line segments, and by implication, the concept of dimension. It would be preferable to construct a purely abstract system based upon elements that are not initially derived from aspects of physical experience, and which do not initially reflect properties of the physical world. It was with this goal that the system presented here was constructed. It rests upon only three primitive concepts, which are not seen as being derived from physical experience. Even though the names that are used are suggestive of physical entities and familiar concepts that is solely to facilitate understanding. Indeed, even some of the familiar terms are used somewhat differently than in traditional systems. The definitions themselves contain very few properties, and no a priori assumptions are made. As such, it is a pure mathematical system.

There are various directions in which such a system could be extended. The one presented here is used, like those of Hellman and Shapiro, as representation of a one-dimensional space, but others might represent spaces of more dimensions and be used for other purposes.

Part 1

We begin with two undefined terms, that is, primitive notions: "corpus" and "in"[1].

1. A *corpus* is an axiomatic object having no assumed properties other than being subject to the relationship "in". We indicate corpora by upper case letters A, B, etc.

2. Between corpora there can obtain an undefined relationship called "in". Between any two corpora, the relationship *in* may or may not obtain (i.e. If A and B are corpora, there are four possibilities: A is in B but B is not in A; B is in A but A is not in B; A is in B and B is in A; A is not in B and B is not in A.)

In is *transitive*: if A is in B and B is in C, then A is in C. It is also *reflexive*: every corpus is in itself - that is, for all A, A is in A. We shall use several expressions for this basic relationship, all of which are equivalent. *Contain* is the verb used to express the relationship of "inness", to mean that one corpus is in another, and $\supset$ is the symbol. It is used freely in either direction: "A is in B" means the same as "B contains A", and the same as "$A \subset B$" or "$B \supset A$".

We shall now proceed to make several definitions:

Definition. Identity of corpora. A is *identical* to B if $\forall$ X, $X \subset A \Rightarrow X \subset B$, & $X \subset B \Rightarrow X \subset A$[2]. We indicate identity by $A \equiv B$.

---

[1] There is nothing in our definitions that says what a corpus is. It could be anything, as long as it satisfies these few conditions. Although we shall be visualising corpora as connected regions, that is, a one dimensional corpus as a line segment; a two dimensional corpus as a disk or arbitrary flat shape, they might be anything. In particular, a single corpus might consist of two or more or even infinite disks or line segments that do not touch each other. When thinking of corpora this way, there is, in principle, no limit to the number of dimensions a corpus may have, but it must have at least one, and it exists in a universe of the same number of dimensions as itself, in which there may be other corpora of the same number of dimensions.
There are other kinds of things that satisfy our description of corpora, such as *sets* and *rules*. Thus the law of gravity on the surface of the earth is "in" the law of gravity at any distance from the centre of the earth, which is "in" the law of universal gravity between any two masses. Many other kinds of things, however, are excluded by the need to satisfy a relationship "in". Thus if corpora were human beings, no relationship such as "son of" "brother of" or "friend of" would satisfy "in".
We shall begin our discussion with general corpora, and then restrict them step by step, until we obtain something having the properties of our intuitive concept of line segments on an endless line.

[2] Intuitively, A is identical to B means that they are two names for the same corpus. We are following the intuitive idea that two corpora are identical if there is nothing in one that is not in the other. That is, not only is A in B, but A fills B up. If X is another corpus that is in B, then X is also in A. This statement is actually stronger than necessary. Stating "if A is in B and B is in A" would have been sufficient. This is implied by the definition we have chosen, since A is in itself and B is in itself. However, this stronger definition will be useful later.

By the definition of identity, it follows that $A \subset B \; \& \sim A \equiv B \Rightarrow \exists X \ni X \subset B \& \sim X \subset A$.[3]

Definition. A and B are *conjunct* if f$\exists X \ni X \subset B \& X \subset A$. We indicate "A and B are conjunct" thus: AoB.

Definition. *Disjunct* is the negation of conjunct. If A and B are not conjunct they are disjunct. That is, there is no corpus that is in both A and B.[4] We indicate "A and B are disjunct" thus: AxB.

It follows directly from this definition that if A is in B or B is in A, then they are conjunct.

Definition. A *space, S,* is a set of corpora $\ni \forall A \& B \in S \; \exists C \in S \ni A \subset C \& B \subset C$.

We indicate spaces by italic uppercase letters.[5]

Definition. *Intersection.* X is the *intersection* of A and B if $X \subset A \& X \subset B \& \forall Y, Y \subset A \& Y \subset B \Rightarrow Y \subset X$. We indicate intersection thus: $A \cap B$. [6]

---

[3] Note that ~ is used without parentheses for negation of a statement or relationship since here is no ambiguity in this notation.

[4] If two corpora are disjunct, there is no difference whether they are "right next to" one another or far apart. Indeed, even when, later on, we define size and distance, distance will have no meaning in itself - only in terms of intervening corpora. Only when there is at least one other corpus between them is there distance between them.

This definition of space is very different from our usual one. Our strategy from here on will be to begin with this very general definition, and proceed to impose conditions that will eventually lead to a space that has the properties of a line, which is to say, agrees with our ordinary concept of a one dimensional space. So too there may be defined spaces that have the properties of two, three and n-dimensional spaces. Again, the important thing here is that we will be able to do this without using a-priori concepts of point or line.

Note also that, if the space is thought of as a line and the corpora as line segments in it, while from 'within' the universe this line may appear straight, if that universe were to be embedded in one of two or more dimensions, there it might not be. Since the one-dimensional universe in which the corpus exists is its whole universe, there is no difference between straight and curved. It might even bend over and cross itself, in which case the meeting of the two parts of the universe would not even exist at either of those places, because the one would be dimensionless in the other.

[6]. Note how this differs from the usual definition of intersection. It is possible for two corpora to be conjunct, yet not have an intersection. There may be several corpora that are in both A and B, without there being any corpus in both A and B that contains them all. Drawing A and B in the usual way is misleading, because it looks as if there is a shape there, but that shape is not part of the system because it has not been defined as a corpus.

Definition. *Convex.* A space S is *convex* if f$\forall$ A $\in S$ & B $\in S \ni$ AoB, $\exists$ X $\ni$ X $\equiv$A$\cap$B.[7]

Definition. *Separable.* A space, S, is *separable* if f$\forall$ A $\in S$ & B $\in S$ & ~B$\subset$A $\exists$ C $\in S \ni$ C$\subset$B & CxA.[8]

Definition. *Complement.* X is the *complement* of A in B if X$\subset$B & XxA & $\forall$ Y$\ni$ Y$\subset$ & YxA, Y$\subset$ X. We indicate the complement of A in B by B-A.[9]

Definition. *Linear.* A space, S, is *linear* if f$\forall$ A $\in S$ & B $\in S$ & ~B$\subset$A, $\exists$ B-A.[10]

Definition. *Union*: If a space, S, is convex and linear, & $\exists$ C $\ni$ C-A $\equiv$ B & C-B $\equiv$ A, then C is called the *union* of A and B. We indicate union thus: A$\cup$B.[11]

Definition. A*djacent*: If a space, S, is convex and linear, then two corpora, A and B, that are elements of S are *adjacent* if $\exists$ C $\ni$ C$\equiv$ A$\cup$B. We indicate adjacent thus: A|B.

It follows that if a space, S, is convex and linear, A $\in S$ & B $\in S$ & AoB & ~A$\subset$B & ~B$\subset$A then A$\equiv$ (A$\cap$B)$\cup$(A-B) & B $\equiv$ (A$\cap$B)$\cup$(B-A)[12]

---

[7] This excludes the possibility of A and B overlapping in more than one place, and there being several corpora that are in both, but without there being any corpus that is in both and contains both of them. That could happen if corpora could have protusions, such as being crescent shaped, hence the choice of the term "convex". Note that here and in future definitions, we place restrictions on the kind of corpora by defining a kind of space. Note also, that we have nowhere stated what corpora are like - in particular, whether they are 'connected' or not. If a corpus could consist of two separate discs, then the restriction of convexity would look very different. However, we shall not concern ourselves with such possibilities because later restrictions will take care of them. For the meanwhile, however, there are indeed several possibilities that could satisfy this.

[8] If A and B are disjunct, this is trivially true. It is significant only in the case where A and B are conjunct. The definition of 'in' only implies that if B is not in A, there exists a C that is in B but not in A. Indeed, B itself satisfies this. That C, however, could still be conjunct with A, because part of A or even all of A could be in C. Separability says that there exists a C that is in B and disjunct from A.

[9]. Complement is the counterpart of intersection.

[10] This doesn't look much like a line, but like convexity, it is named for that which motivated it and towards which it is directed. These properties will eventually give us a space that behaves like a line.

[11] We are now beginning to define composition and addition. It is clear why it has been necessary to do it this way, given the nature of the definitions with which we started.

Part 2

Definition. A space, *S,* is *upwardly nested* if f$\forall A \in S \exists B \in S \ni A \subset B \,\&\, \sim A \equiv B$.

Definition. A space, *S*, is *downwardly nested* if f$\forall A \in S \exists B \in S \ni B \subset A \,\&\, \sim A \equiv B$.[13]

Definition. A space, *S*, is *compact* if f$\forall A \in S \exists B \in S \,\&\, C \in S \ni \sim A \equiv B \,\&\, \sim A \equiv C \,\&\, B \cap C \equiv A$.[14]

Definition. A space, *S*, is *spanable* if f$\forall A \in S \,\&\, B \in S \,\&\, A|B \exists C \in S \,\&\, D \in S \,\&\, E \in S \ni C \subset A \cup B \,\&\, CoA \,\&\, CoB \,\&\, D \subset A \,\&\, DxC \,\&\, E \subset B \,\&\, ExC$.

Definition. A space is *bi-directional* if f$\forall A \in S \exists B \in S \,\&\, C \in S \ni AoB \,\&\, AoC \,\&\, \sim C \subset A \,\&\, \sim B \subset A \,\&\, CxB$.[15]

Part 3

We shall now introduce a third undefined term, *length.* Length is a scalar value identified with a corpus. It can have any real value greater than 0. Every corpus has a length. Several corpora can have the same length.[16] We denote the length of corpus A in two ways, by [A] and by lower-case characters, a.

---

[12] What we now have is a space such that for any A and B that are conjunct, but neither of which is in the other, there exist C, D and E such that C is the intersection of A and B, D is the complement of A in B, and E is the complement of B in A.

[13] Note that even though both of these kinds of spaces have infinite corpora, their elements do not necessarily become very small or very large. (Of course, we have not yet defined any kind of measurement, but even when eventually we do, the corpora could, for example, get smaller and smaller but limit at one half, or larger and larger but limit at two)

[14] Compactness is important because it guarantees the possibility of adding to a corpus. Without it, a corpus could be all by itself. Even though it would have smaller ones inside it and larger ones around it, it wouldn't necessarily have anything adjacent to it. This is saying that the space is tightly packed with corpora that can be united with one another to make larger corpora.

[15] Without this property, the space might have an infinity of corpora each including more than the last, but each more inclusive one would contain all of the less inclusive ones. This way, it is possible to have some that are not contained in others, which corresponds to our image of going off in two directions. The expression "bi-directional" is not intended to mean exactly two. Two is just a minimum. There could be any number or infinitely many.

[16] This takes the place of the conventional concept of length, but is necessarily radically different, because it is not defined by the distances between points.

Definition. *Equal* If two corpora have the same length, we say they are *equal* to one another. We indicate equality in the usual way, by =.

Length has the following qualities:

1. If $A \subset B$ and $\sim A \equiv B$, then $[A] < [B]$. (It need not be stated that if $A \equiv B$ then $[A] = [B]$, since if they are identical they are just different names for the same corpus)

2. $[A \cup B] = [A] + [B]$.[17]

Definition. We now arbitrarily choose one specific corpus as a standard of length and assign it a value of 1. We then compare other corpora to it to assigning them lengths in the following way for rational numbers[18]: The length of a corpus A is n if it is the union of n corpora that have length 1. The length of a corpus A is n/m if there exists a corpus C which is the union of m corpora that have length [A] and is also the union of n corpora that have length 1.[19]

Definition. We now arbitrarily choose two adjacent corpora, which we call P and N, and call any corpus that is conjunct with P but not with N *positive*, and any corpus that is conjunct with N but not with P *negative*.

Definition. *Positive Cartesian, Negative Cartesian:* A corpus, A, is *Positive Cartesian* if one of the following apply: either 1. $P \subset A$ & AxN; or 2. $\exists B \ni B \subset N \cup P$ & BoN & BoP& $A \equiv B \cap P$.

Negative Cartesian is defined similarly. We indicate positive and negative Cartesian by A>0 and A<0 respectively (Since A is a corpus, not a scalar number, this should not cause any confusion). [20]

Definition. *Cartesian Space*: A space, *S*, is *Cartesian* if f$\forall$ n $\in R \exists$ A $\in S$& B$\in S \ni$ A>0 & B<0 & [A] = n & [B] = n.

---

[17] The second property is stronger than the first and implies it. However, since union is only defined for two corpora that are adjacent, not for corpora that are conjunct, there can be spaces in which there is no way of uniting corpora, so the second property would not apply yet the first would.

[18] I have not decided what to do about irrational numbers. This question can be dealt with in the future.

[19] All these corpora must actually exist to be able to say what the length of A is. Otherwise, all we can say is that A has some length, and that length is greater than the length of any corpus that is in A, and less than the length of any corpus that contains A.

[20] In this way we have created something that functions like an origin, and defined corpora beginning at it and extending in one direction as "positive" and in the other as "negative".

Definition. *Measurably compact.* A space, *S*, is *measurably compact* if f$\forall$ A $\in$ *S*& n $\in$*R*& m $\in$*R*$\exists$ B $\in$ *S* & C$\in$ *S* $\ni$ A $\equiv$ (B$\cap$C) & [C-A] = n & [B-A] = m.[21]

Definition. *Measurably bi-directional*: A space, *S*, is *measurably bi-directional* if f$\forall$ A $\in$ *S* & B $\in$ *S* & n $\in$*R* $\ni$ AoB & $\sim$ B$\subset$A $\exists$ C $\in$ *S* $\ni$ CoB & $\sim$ C$\subset$B & CxA & [C-B] = n.[22]

Definition. A space is *measurably spanable* if f$\forall$ A $\in$ *S* & B $\in$ *S* & AoB & $\sim$ B$\subset$A & $\sim$ A$\subset$B &$\forall$ n $\in$*R* &$\forall$ m$\in$*R* $\ni$0<m<[A$\cap$B] & 0<n<[B-A] $\exists$ C $\in$ *S* $\ni$C$\subset$B & [C$\cap$A] = m & [C-A] = n.[23]

Definition. *Connected*: A space, *S*, is *connected* if f$\forall$ A $\in$ *S* & B $\in$ *S*$\ni$ B|A, &$\forall$ m $\in$*R*$\ni$ m > [B] $\exists$ C $\in$*S*$\ni$ [C] = m & C|A, & C$\supset$B.[24]

A space satisfying these conditions can be used as we conventionally use a line, scalars being represented by corpora rather than points, and functions mapped from one space to another. A function such as y = f(x) looks the same in this as it does in the conventional system, even though x and y are defined not in terms of end points but in terms of corpora of those lengths. A series of corpora whose lengths go to 0 can be produced anywhere, so we can express limits using epsilon and delta. We can also now define points, not as undefined axiomatic entities, but as limits of nested corpora spanning two adjacent corpora.

As we have already said, the usual system of points and lines works perfectly well for all practical purposes. The reason for constructing this new system is to prove that such a foundation is possible, that dimensionless points and combinations of entities of different numbers of dimensions are not necessary to represent the physical world. This system is appropriate for representing any continuous scalar such as time or any of the three dimensions of space, and for representing relationships between continuous scalars. (For representing discrete

[21] This says that there are adjacent to A corpora of any size, large or small.

[22] This says that B has a property corresponding to "two sides" or "two directions", and that furthermore, in one of those directions, the direction of C, there are corpora adjacent to B having lengths equal to all rational numbers. Although not stated explicitly, it is obvious that had we started with C in place of A, we could have gone in another direction.

[23] We could have done without the earlier definition of spanable, and just call this one "spanable".

[24] This last condition eliminates a possibility that we have not bothered to address up until now, that a corpus look not like an interval, but perhaps like two intervals separated by some distance, perhaps very large, or even of many or perhaps an infinite number of such intervals. This says that on any side of the corpus at which it is possible to place an adjacent corpus, no matter how small, it is also possible to place unlimited larger corpora reaching out across the space, and none of them will ever meet another part of the original corpus.

entities such as animals of human beings, or the number of times an event occurs, such a system would, of course, be neither necessary nor appropriate.)

References

Aristotle [1960]: *The Physics.* trans. P. H. Wicksteed and F. M. Cornford, Cambridge, MA: Harvard University Press.

Bolzano, Bernard [1950]: *Paradoxes of the Infinite*, trans. F. Prihonsky, New Haven: Yale Press.

Cantor, Georg [1955]: *Contributions to the Founding of the Theory of Transfinite Numbers,* trans. Philip E. B. Jordain, New York: Dover Publications.

Descartes, René [1886]: *La Géométrie*, [Geometry] Paris: A Hermann Librairie Scientifique.

Glazebrook, Trish [2001]: 'Zeno against mathematical physics', *Journal of the History of Ideas* 62(2) 193-210.

Hellman, Geoffrey, and Stewart Shapiro [2012]: 'Towards a point-free account of the continuous', *Iyyun: The Jerusalem Philosophical Quarterly* 61, 263-278.

- [2013]: 'The classical continuum without points', *Review of Symbolic Logic* 6, 488-512.

Linnebo, Øystein, Stewart Shapiro, and Geoffrey Hellman [2016]: 'Aristotelian Continua', *Philosophia Mathematica* 24(2) 214-246.

Owen, G. E. L. [1970]: 'Zeno and the mathematicians', in: W. C. Salmon, ed., *Zeno's Paradoxes*, New York: Bobbs-Merrill, pp. 139-163.

Papa-Grimaldi, Alba [1996]: 'Why mathematical solutions to Zeno's paradoxes miss the point: Zeno's one and many relation and Parmenides' prohibition', *The Review of Metaphysics* 50(2), 299-314.

Rodych, Victor [2011]: 'Wittgenstein's Philosophy of Mathematics', *The Stanford Encyclopedia of Philosophy* (Summer 2011 Edition), Edward N. Zalta(ed.).

Tannery, Paul [1885]: 'Le concept scientifique du continu: Zénon d'Elée et Georg Cantor,' [The scientific concept of the continuum: Zeno of Elea and Georg Cantor] *Revue Philosophique de la France et de l'Aetranger*, 20(2), 385-410.

INDEX

E

F

G

H

I

P

Q

R

S

T

U

V

W

Y